Charles Kau

THE ARCHZEOLOGICAL COLLECTION OF THE UNITED STATES NATIONAL MUSEUM

Charles Kau

THE ARCHZEOLOGICAL COLLECTION OF THE UNITED STATES NATIONAL MUSEUM

ISBN/EAN: 9783742894199

Manufactured in Europe, USA, Canada, Australia, Japa

Cover: Foto ©berggeist007 / pixelio.de

Manufactured and distributed by brebook publishing software (www.brebook.com)

Charles Kau

THE ARCHZEOLOGICAL COLLECTION OF THE UNITED STATES NATIONAL MUSEUM

THE ARCHZEOLOGICAL COLLECTION OF THE UNITED STATES NATIONAL MUSEUM

CHARLES KAU

ADVERTISEMENT.

THE author of this work was entrusted with the classification of the Smithsonian Archaeological and Ethnological Collections before they were transferred to Philadelphia, to form a part of the United States Government Representation at the International Exhibition of 1876.

While thus engaged, he found time to prepare the following account, which, though far from being exhaustive, will at least serve to show what ample means the National Museum presents for the study of North American archæology.

JOSEPH HENRY,
Secretary S. I.

SMITHSONIAN INSTITUTION,
Washington, August, 1876.

CONTENTS.

	PAGE
INTRODUCTION	1
STONE	7
A. Flaked and chipped Stone	7
1. Raw Material	7
2. Irregular Flakes of Flint, Obsidian, etc., produced by a single blow	8
3. Two-edged narrow Flakes of Obsidian and prismatic Cores or Nuclei, from which such Flakes have been detached by pressure	8
4. Pieces of Flint, Quartz, Obsidian, etc., roughly flaked, and either representing rude tools, or designed to be wrought into more regular forms.	
—Unfinished Arrow and Spear-heads	8
5. Arrow-heads	8
6. Spear-heads	10
7. Perforators	12
8. Scrapers	13
9. Cutting and Sawing Implements	13
10. Dagger-shaped Implements	14
11. Leaf-shaped Implements	15
12. Large flat Implements of silicious material, usually ovoid in shape, and sharp around the circumference (Digging Tools)	16
13. Large flat Implements, mostly of oval outline, but truncated and laterally notched at the end opposite the working edge (Digging Tools)	16
14. Wedge or Celt-shaped Implements	17
B. Pecked, ground and polished Stone	17
1. Wedges or Celts	17
2. Chisels	18
3. Gouges	18
4. Adzes	19
5. Grooved Axes	19
6. Hammers	21
7. Drilled Ceremonial Weapons	23
8. Cutting Tools	24
9. Scraper and Spade-like Implements	25
10. Pendants and Sinkers	26
11. Discoidal Stones and Implements of kindred Shape	28
12. Pierced Tablets and Boat-shaped Articles	32
13. Stones used in Grinding and Polishing	34

CONTENTS.

	PAGE
14. Vessels	36
15. Mortars	38
16. Pestles	41
17. Tubes	43
18. Pipes	45
19. Ornaments	51
20. Sculptures	54
COPPER	59
BONE AND HORN	63
SHELLS	66
CLAY	73
WOOD	88
SUPPLEMENT	90
APPENDIX I.	93
The Aboriginal Modes of hafting Stone and Bone Implements	93
APPENDIX II.	97
System adopted in arranging the Smithsonian Collection illustrative of North American Ethnology	97
I. Man, Objects relating to,	97
II. Culture, Objects relating to,	97
INDEX	103

LIST OF ILLUSTRATIONS.

Objects of flaked and chipped Stone.

			PAGE
Fig.	1.	Flint Knife in wooden handle; Pai-Utes	2
Fig.	2.	Obsidian Knife; Mexico	6
Fig.	3.	Obsidian Core; Mexico	8
Fig.	4.	Arrow-head, leaf-shaped; semi-opal; California	9
Fig.	5.	Arrow-head, convex-sided, truncated base; obsidian; Mexico . .	9
Fig.	6.	Arrow-head, triangular; jasper; New York	9
Fig.	7.	Arrow-head, indented base; jasper; Oregon	9
Fig.	8.	Arrow-head, notched at the sides near the base; jaspery agate; Texas	9
Fig.	9.	Arrow-head, notched at the sides; hornstone; Tennessee . . .	9
Fig.	10.	Arrow-head, notched at the sides; hornstone; Ohio	9
Fig.	11.	Arrow-head, stemmed; flint; Ohio	9
Fig.	12.	Arrow-head, stemmed; hornstone; Pennsylvania	9
Fig.	13.	Arrow-head, stemmed; silicified wood; Ohio	9
Fig.	14.	Arrow-head, stemmed; hornstone; Ohio	9
Fig.	15.	Arrow-head, stemmed; quartz; District of Columbia . . .	9
Fig.	16.	Arrow-head, stemmed; flint; Tennessee	9
Fig.	17.	Arrow-head, stemmed; hornstone; Tennessee	9
Fig.	18.	Arrow-head, stemmed; hornstone; Tennessee	9
Fig.	19.	Arrow-head, barbed and stemmed; hornstone; Tennessee . . .	9
Fig.	20.	Arrow-head, barbed and stemmed; semi-opal; Oregon . . .	9
Fig.	21.	Arrow-head, barbed and stemmed; semi-opal; Oregon . . .	9
Fig.	22.	Arrow-head, beveled on opposite sides; flint; Tennessee . .	9
Fig.	23.	Arrow-head, serrated; flint; Oregon	9
Fig.	24.	Arrow-head, serrated; jasper; Louisiana	9
Fig.	25.	Arrow-head, stem bifurcated; jasper; Tennessee	9
Fig.	26.	Spear-head, convex-sided, base truncated; chalcedony; Tennessee	11
Fig.	27.	Spear-head, concave base; jasper; California	11
Fig.	28.	Spear-head, notched near the base; flint; Kentucky . . .	11
Fig.	29.	Spear-head, stemmed; quartz schist; Pennsylvania . . .	11
Fig.	30.	Spear-head, stemmed; flint; New York	11
Fig.	31.	Spear-head, barbed and stemmed; milky quartz; Louisiana . .	11
Fig.	31a.	Spear-head, with several notches near the base; jasper; Maine .	11
Fig.	32.	Perforator, triangular; jasper; Ohio	12
Fig.	33.	Perforator, slender, expanding opposite the point; jasper; Oregon	12
Fig.	34.	Perforator, slender, broad base; flint; Missouri	12
Fig.	35.	Perforator, indented base; flint; Ohio	12
Fig.	36.	Perforator, stemmed; hornstone; Tennessee	12
Fig.	37.	Perforator, elongated leaf-shape; semi-opal; California . . .	12
Fig.	38.	Scraper, beveled from one side; flint; Texas	13
Fig.	39.	Scraper, beveled from one side, stemmed; hornstone; Ohio . .	13
Fig.	40.	Scraper, made of a broken arrow-head; jasper; Ohio . . .	13
Fig.	41.	Scraper, disc-shaped, chipped all around; chalcedony; Texas . .	13

(vii)

viii   LIST OF ILLUSTRATIONS.

		PAGE
Fig. 42.	Flake (knife); jasper; Kentucky	14
Fig. 43.	Cutting Tool, with stems opposite the convex edge; flint; California	14
Fig. 44.	Cutting Tool, sickle-shaped; jasper; California	14
Fig. 45.	Cutting Tool, crescent-shaped, lydite; Pennsylvania	14
Fig. 46.	Cutting Tool, arrow-head-shaped; jasper; Tennessee	14
Fig. 47.	Cutting Tool, arrow-head-shaped; semi-opal; Georgia	14
Fig. 48.	Cutting Tool, oval; hornstone; Maine	14
Fig. 49.	Dagger; flint; Alabama	15
Fig. 50.	Leaf-shaped Implement, pointed at one end; flint; Ohio	15
Fig. 51.	Leaf-shaped Implement, broad, pointed at both ends; jasper; Louisiana	15
Fig. 52.	Leaf-shaped Implement, slender, pointed at both ends; flint; Ohio	15
Fig. 53.	Leaf-shaped Implement, large, pointed at one end; hornstone; Illinois	15
Fig. 54.	Large Implement of oval outline (digging tool); fine-grained quartzite; Tennessee	16
Fig. 54a.	Large Implement, with broad convex working edge, truncated at the opposite end (digging tool); fine-grained quartzite; Illinois	16
Fig. 55.	Large Implement, oval, truncated and laterally notched at the end opposite the working edge (digging tool); fine-grained quartzite; Illinois	16

Objects of pecked, ground and polished Stone.

Fig. 56.	Celt, small; hematite; Ohio	17
Fig. 57.	Celt, four-sided; greenstone; Indiana	17
Fig. 58.	Celt, broad butt-end; syenite; Illinois	17
Fig. 59.	Celt, battered butt-end; greenstone; Tennessee	17
Fig. 60.	Celt, tapering butt-end; indurated chlorite slate; Tennessee	17
Fig. 61.	Celt, expanding cutting edge; cast; Louisiana	17
Fig. 62.	Celt, terminating in a handle opposite the edge; greenstone; North Carolina	17
Fig. 63.	Chisel, round in the cross-section; diorite; Ohio	19
Fig. 64.	Chisel, four-sided; lydite; New York	19
Fig. 65.	Chisel, with handle; basaltic stone; Unalaska Island	19
Fig. 66.	Gouge, partly hollowed; hornstone; New York	19
Fig. 67.	Gouge, entirely hollowed; argillite; Pennsylvania	19
Fig. 68.	Adze, gouge-like; greenstone; Massachusetts	19
Fig. 69.	Adze; hornstone; British Columbia	19
Fig. 70.	Hafted Adze; greenstone; Oregon	19
Fig. 71.	Hafted Adze; serpentine; Northwest Coast	19
Fig. 72.	Grooved Axe, groove passing entirely around; greenstone; Massachusetts	20
Fig. 73.	Grooved Axe, one side flat; greenstone; Arizona	20
Fig. 74.	Grooved Axe, groove bounded by ridges; greenstone, South Carolina	20
Fig. 75.	Grooved Axe, unusually long; cast; Wisconsin	20
Fig. 76.	Grooved Axe, four-sided; greenstone; Alaska	20
Fig. 77.	Grooved Axe, narrow cutting edge; graywacke; Pennsylvania	20
Fig. 78.	Maul, grooved; granite; Colorado	20
Fig. 79.	Hafted Maul; quartzite; Assineboins	20
Fig. 80.	Hammer-stone, with a depression on each side; quartzite; New York	20
Fig. 81.	Hammer-stone, with a depression on each side; quartzite; Pennsylvania	20
Fig. 82.	Hammer-stone; flint; Ohio	20
Fig. 83.	Drilled Ceremonial Weapon, pick-shaped; serpentine; Virginia	23
Fig. 84.	Drilled Ceremonial Weapon, pick-shaped; serpentine; Pennsylvania	23
Fig. 85.	Drilled Ceremonial Weapon, pick-shaped; slate; Wisconsin	23
Fig. 86.	Drilled Ceremonial Weapon, pick-shaped; slate; Indiana	23
Fig. 87.	Drilled Ceremonial Weapon, much curved, ends expanding; slate; Pennsylvania	23

LIST OF ILLUSTRATIONS.   ix

		PAGE
Fig. 88.	Drilled Ceremonial Weapon, parts corresponding to edges strongly curved; cast; Louisiana.	23
Fig. 89.	Drilled Ceremonial Weapon, axe-shaped; cast; Wisconsin	23
Fig. 90.	Drilled Ceremonial Weapon (fragment), bifurcated; slate; Indiana	23
Fig. 91.	Drilled Ceremonial Weapon, inwardly curved sides; translucent ferruginous quartz; Indiana	23
Fig. 92.	Drilled Ceremonial Weapon, crooked, blunt at one end; slate; Indiana	23
Fig. 93.	Cutting Tool, semi-lunar, pierced; slate; Pennsylvania	25
Fig. 94.	Cutting Tool, with lateral tang; hard red shale; Pennsylvania	25
Fig. 95.	Cutting Tool, with long handle; cast; Indiana	25
Fig. 96.	Scraper-like Tool, semi-circular edge, pierced; greenstone; Kentucky	25
Fig. 97.	Scraper-like Tool, slightly convex edge; cast; Arkansas	25
Fig. 98.	Large Tool, with convex edge, the opposite end forming a bifurcation; cast; South Carolina	25
Fig. 99.	Spade-like Implement; cast; South Carolina	25
Fig. 100.	Pendant, nearly pear-shaped; hornblende rock; Ohio	27
Fig. 101.	Pendant, nearly pear-shaped, grooved at one end; hematite; Tennessee	27
Fig. 102.	Pendant, elongated pear-shape, perforated at one end; amygdaloid; Arkansas	27
Fig. 103.	Pendant, swelling in the middle, grooved at one end; greenstone; Ohio	27
Fig. 104.	Pendant, double conoid-shape; greenstone; California	27
Fig. 105.	Sinker, expanding at the upper end and terminating in a knob; quartzite; Massachusetts	27
Fig. 106.	Sinker, pear-shaped, with a knob at the narrower extremity; greenstone; Massachusetts	27
Fig. 107.	Sinker, globular, encircled by a groove; granite; Rhode Island	27
Fig. 108.	Sinker, nearly globular, encircled by a groove; potstone; Georgia	27
Fig. 109.	Sinker, encircled by two grooves intersecting each other; talcose slate; Rhode Island	27
Fig. 110.	Sinker (?), of elongated shape, encircled by a groove and ornamented with incised lines; sandstone; Oregon	27
Fig. 111.	Sinker, notched flat pebble; quartzite; Pennsylvania	27
Fig. 112.	Sinker, notched; graywacke; New York	27
Fig. 113.	Sinker, notched; quartzite; Pennsylvania	27
Fig. 114.	Sinker, pierced in the centre; micaceous slate; California	27
Fig. 115.	Sinker, pierced obliquely near the circumference; sandstone; Ohio	27
Fig. 116.	Discoidal Stone, concave on both sides; ferruginous quartz; Tennessee	29
Fig. 117.	Discoidal Stone, concave on both sides; ferruginous quartz; Tennessee	29
Fig. 118.	Discoidal Stone, concavities on both sides; greenstone; Illinois	29
Fig. 119.	Discoidal Stone, concave on both sides and perforated in the centre; cast; Ohio	29
Fig. 120.	Discoidal Stone, concave on both sides and perforated; quartzite; Ohio,	29
Fig. 121.	Discoidal Stone, flat on both sides; quartzose stone; Georgia	29
Fig. 122.	Discoidal Stone, small, flat on both sides; argillite; Pennsylvania	29
Fig. 123.	Discoidal Stone, lenticular; ferruginous quartz; Texas	29
Fig. 124.	Club-head-shaped Stone; hornblende rock; California	31
Fig. 125.	Club-head-shaped Stone; greenstone; California	31
Fig. 126.	Club-head-shaped Stone; serpentine; California	31
Fig. 127.	Pierced Tablet, longer sides convex; slate; New York	33
Fig. 128.	Pierced Tablet, longer sides concave; slate; Pennsylvania;	33
Fig. 129.	Pierced Tablet, four sides curved inwardly; cast; Louisiana	33
Fig. 130.	Pierced Tablet, lozenge-shaped; slate; Tennessee	33
Fig. 131.	Pierced Tablet, pentagonal, longer sides concave; slate; Tennessee	33
Fig. 132.	Pierced Tablet, longer sides concave, shorter sides convex; slate; Tennessee	33
Fig. 133.	Irregular flat Stone, showing several perforations; potstone; Pennsylvania	33
Fig. 134.	Boat-shaped Object, solid, two perforations; slate; Ohio	33
Fig. 135.	Boat-shaped Object, hollowed, a perforation near each end; greenstone; Kentucky	33
Fig. 136.	Oval Pebble, showing furrows radiating toward the circumference; quartzose rock; New Jersey	34
Fig. 137.	Arrow-shaft-grinder; chlorite slate; Massachusetts	34

LIST OF ILLUSTRATIONS.

		PAGE
Fig. 138.	Arrow-shaft-grinder; compact chlorite; Mexico	84
Fig. 139.	Arrow-shaft-grinder; hornblende rock; Southern Utah	84
Fig. 140.	Polisher, oval, with truncated ends; quartzose rock; Indiana	85
Fig. 141.	Polisher, rhomboidal in outline; cast; Louisiana	85
Fig. 142.	Polisher, club-shaped; lydite; Pennsylvania	85
Fig. 143.	Vessel of elongated shape, with projections at the narrower ends; potstone; Massachusetts	86
Fig. 144.	Bowl; potstone; Wyoming	86
Fig. 145.	Large Vessel, nearly globular, with a narrow aperture encircled by a raised rim; potstone; California	36
Fig. 146.	Large bowl-shaped Vessel; potstone; California	36
Fig. 147.	Boat-shaped Vessel; potstone; California	86
Fig. 148.	Bowl; serpentine; California	36
Fig. 149.	Cup; sandstone; California	36
Fig. 150.	Round Plate, ornamented; graywacke; Alabama	87
Fig. 151.	Rectangular Plate, ornamented; graywacke; Alabama	87
Fig. 152.	Pierced Plate; potstone; California	37
Fig. 153.	Mortar, hollowed boulder; sandstone; California	39
Fig. 154.	Mortar, hollowed boulder; sandstone; California	80
Fig. 155.	Mortar, with projections on two sides; sandstone; California	80
Fig. 156.	Large Mortar, tapering toward the bottom; sandstone; California	39
Fig. 157.	Small Mortar, tapering toward the bottom, ornamented; sandstone; California	39
Fig. 158.	"Metate;" sandstone; Utah	80
Fig. 159.	Slab of granite with sandstone Rubber; Navajo Indians	39
Fig. 160.	Slab of sandstone, showing irregular depressions ("nut-stone"); Pennsylvania	40
Fig. 160a.	Slab of sandstone, bearing cup-shaped cavities; Kentucky	40
Fig. 161.	Pestle of conical form; syenite; California	42
Fig. 162.	Pestle, terminating in a knob at the upper end; sandstone; California	42
Fig. 163.	Pestle, with a small knob at the upper extremity; sandstone; California	42
Fig. 164.	Pestle, showing a ring-like ridge below the tapering upper end; sandstone; California	42
Fig. 165.	Pestle, with a knob-like expanse at the lower extremity; amygdaloid; California	42
Fig. 166.	Pestle of cylindrical form; sandstone; Rhode Island	42
Fig. 167.	Cylindrical Pestle, tapering abruptly at the upper end; compact greenstone; Alaska	42
Fig. 168.	Pestle of conical form, expanding base; greenstone; Pennsylvania	42
Fig. 169.	Pestle, with expanding base and truncated upper end; syenite; Ohio	42
Fig. 170.	Pestle, showing an expanding base and a circular ridge below the tapering upper end; greenstone; British Columbia	42
Fig. 171.	Pestle, with expanding base and similarly shaped upper extremity; silicious stone; Washington Territory	42
Fig. 172.	Pestle, exhibiting a horizontal handle which terminates in round plates; greenstone; Alaska	42
Fig. 173.	Disc-shaped Stone, probably employed for grinding purposes; greenstone; Georgia	42
Fig. 174.	Muller, conoid-shaped; greenstone; Ohio	42
Fig. 175.	Cylindrical Tube; steatite; Tennessee	44
Fig. 176.	Tube, encircled in the middle by a raised ring; chlorite; Tennessee	44
Fig. 177.	Pipe, showing a round expanding bowl on a slightly curved base ("mound-pipe"); cast; Ohio	47
Fig. 178.	Pipe, with a bowl in the shape of a human head; cast; Ohio	47
Fig. 179.	Pipe, carved in imitation of a beaver; cast; Ohio	47
Fig. 180.	Pipe, in the shape of an otter with a fish in its mouth; cast; Ohio	47
Fig. 181.	Pipe, representing a heron; cast; Ohio	47
Fig. 182.	Pipe, showing the body of a bird with a human head; cast; Ohio	47
Fig. 183.	Calumet-pipe; human figure with a snake coiled around its neck; cast; Ohio	47
Fig. 184.	Calumet-pipe, fashioned in imitation of a canine animal (wolf?); cast; Ohio	47

LIST OF ILLUSTRATIONS.   xi

		PAGE
Fig. 185.	Calumet-pipe, in the form of a bird (eagle ?); potstone; Kentucky	48
Fig. 186.	Four-sided Pipe, with lateral stem-hole; argillite; Pennsylvania	49
Fig. 187.	Pipe, formed in imitation of a loon, stem-hole; serpentine; West Virginia	49
Fig. 188.	Pipe, representing a parrot, stem-hole; argillite; New York	49
Fig. 189.	Pipe, barrel-shaped bowl, stem hole; argillite; Ohio	49
Fig. 190.	Pipe, with a high bowl rising from a flat base; chlorite; Virginia	49
Fig. 191.	Pipe, in which the connection of the bowl with the neck forms a curve; serpentine; New York	49
Fig. 192.	Pipe with a long stem, lizard-shaped; steatite; Pennsylvania	49
Fig. 193.	Four-sided Pipe, with a neck for the stem; potstone; North Carolina	49
Fig. 194.	Pipe, with a quadrilateral rim showing a human head at each corner; serpentine; Texas	49
Fig. 195.	Pipe, remarkable for a low, broad-rimmed bowl rising from a thick base; limestone; Kentucky	49
Fig. 196.	Clay Pipe, with a neck for a stem; Georgia	49
Fig. 197.	Conoid-shaped Pipe; serpentine; California	49
Fig. 198.	Clay Pipe, representing a coiled snake (fragment); New York	49
Fig. 199.	Clay Pipe, in the shape of a raven's (?) head (fragment); New York	49
Fig. 200.	Bead, globular, compressed on two sides; serpentine; California	52
Fig. 201.	Bead, four-sided; potstone; Pennsylvania	52
Fig. 202.	Bead, with notched circumference; potstone; Pennsylvania	52
Fig. 203.	Tube (bead); silicious stone; Mississippi	52
Fig. 204.	Drilled Ornament, shaped like a compressed slender pyramid; Catlinite; New York	52
Fig. 205.	Gorget, four-sided, with an ornamental border of dotted triangles; trap rock; Connecticut	52
Fig. 206.	Flat oval Pebble, pierced for suspension, and ornamented with incised lines; sandstone; Rhode Island	52
Fig. 207.	Flat oval Pebble, pierced for suspension; sandstone; Pennsylvania	52
Fig. 208.	Shell of hematite, pierced and notched ("record"); Virginia	52
Fig. 209.	Heart-shaped Ornament; argillaceous slate; Ohio	52
Fig. 210.	Bird-shaped Object (amulet?), pierced with an oblique hole at each end of the base; slate; Pennsylvania	52
Fig. 211.	Amulet (?). Though not presenting the shape of a bird, this object bears some analogy to the original of Fig. 210, being pierced with an oblique hole at each extremity of the flat base; slate; Ohio	52
Fig. 212.	Object resembling a cylinder with an inwardly curved side-surface; this surface bears incised ornamental lines; argillaceous sandstone; Kentucky	52
Fig. 213.	Ring, deeply grooved around the circumference and pierced with eight equidistant small holes radiating from the centre; cast; Ohio	52
Fig. 214.	Massive grooved Ring; potstone; Pennsylvania	52
Fig. 215.	Sculptured human Figure, body indistinct; crystalline limestone; Tennessee	53
Fig. 216.	Sculptured human Head; limestone; unknown where found	53
Fig. 217.	Carved human Head; ferruginous stone; Ohio	55
Fig. 218.	Sculpture of a human Head with an elaborate head-dress; volcanic rock; Mexico	55
Fig. 219.	Carving of a human Figure, head only distinctly represented; slate; Mexico	55
Fig. 220.	Carving of a squatting human Figure, neck pierced; alabaster; Mexico	55
Fig. 221.	Carving of a human Skull, pierced; silicified wood; Yucatan	55
Fig. 222.	Sculptured Foot-track; sandstone; Missouri Valley	57
Fig. 223.	Sculptured Foot-track; quartzite; Missouri	57

Objects of Copper.

Fig. 224.	Celt, with adhering pieces of charcoal and cinders; Kentucky	61
Fig. 225.	Axe-shaped Object, terminating at the broader end in lateral curved appendages; Kentucky	61
Fig. 226.	Celt; Tennessee	61

LIST OF ILLUSTRATIONS.

		PAGE
Fig. 227.	Gouge-like Chisel; New York	61
Fig. 228.	Spear-head, with tapering stem; Lake Superior District	61
Fig. 229.	Spear-head, with truncated stem; Vermont	61
Fig. 230.	Crescent-shaped Cutting Tool; Wisconsin	61
Fig. 231.	Awl, inserted in a bone handle; Tennessee	61
Fig. 232.	Sinker; Ohio	61
Fig. 233.	Bead; Ohio	61
Fig. 234.	Tube of copper sheet; Rhode Island	61
Fig. 235.	Spool-shaped Object; Tennessee	61

Objects of Bone.

Fig. 236.	Perforator; Alaska	64
Fig. 237.	Perforator; Kentucky	64
Fig. 238.	Perforator; Kentucky	64
Fig. 239.	Needle, showing two grooves instead of an eye; California	64
Fig. 240.	Harpoon-head, with a hole for attachment; Michigan	64
Fig. 241.	Harpoon-head; perforated; Alaska	64
Fig. 242.	Fish-hook; California	64
Fig. 243.	Whistle, made of a hollow bone; California	64
Fig. 244.	Whistle; California	64
Fig. 245.	Cup, made of a vertebra of a cetacean; California	64
Fig. 246.	Bear's Tooth, drilled; New York	64
Fig. 247.	Bear's Tooth, drilled and polished; Alaska	64
Fig. 248.	Drilled Claw of the grizzly bear; Rocky Mountains; recent	64
Fig. 249.	Drilled Claw of the panther; California	64
Fig. 250.	Drilled Ornament (?), made of the epiphysis of some animal; Kentucky	64
Fig. 251.	Worked hollow Bone; California	64

Objects made of Shells.

Fig. 252.	Bystcon perversum, transformed into a Vessel; Indiana	67
Fig. 253.	Spoon made of a Unio-shell; Kentucky	67
Fig. 254.	Celt-shaped Tool; Florida	67
Fig. 255.	Celt-shaped Tool; Kentucky	67
Fig. 256.	Fish-hook, made of Haliotis-shell; California	67
Fig. 257.	Strombus pugilis, pierced; Florida	69
Fig. 258.	Unio, pierced; Tennessee	69
Fig. 259.	Oliella biplicata, truncated at the apex; California	69
Fig. 260.	Oliva literata, apex removed; Florida	69
Fig. 261.	Pecten concentricus, pierced; Florida	69
Fig. 262.	Cylindrical Bead; California	69
Fig. 263.	Cylindrical Bead; California	69
Fig. 264.	Cylindrical Bead; California	69
Fig. 265.	Prismatic Bead; California	69
Fig. 266.	Bead, made from a columella; Georgia	69
Fig. 267.	Bead, tapering at both ends, showing a part of the columellar groove; California	69
Fig. 268.	Pin-shaped Object; Florida	69
Fig. 269.	String of Wampum, composed of white and violet beads; Upper Missouri	69
Fig. 270.	Disc of Haliotis-shell, with five perforations and ornamented border; California	69
Fig. 271.	Disc of Haliotis-shell, showing five perforations; California	69
Fig. 272.	Gorget, ornamented with an incised design; Tennessee	69
Fig. 273.	Gorget; unornamented; Kentucky	69

LIST OF ILLUSTRATIONS.   xiii

		PAGE
Fig. 274.	Pear-shaped Pendant, grooved at the upper end; New York	69
Fig. 275.	Ring-shaped Pendant of *Haliotis*-shell; California	69
Fig. 276.	Crescent-shaped Ornament, pierced at both ends; California	69
Fig. 277.	Object of *Haliotis* shell, irregular in outline, pierced with four holes, and ornamented along the border; California	64
Fig. 278.	Object of *Haliotis*-shell, irregular in shape, pierced with one hole; California	69
Fig. 279.	Object, made of *Lucapina crenulata*, nearly oval in outline, and showing a large oval aperture in the middle; California	69

Objects of Clay.

Fig. 280.	Bowl, with four projections at the rim; Tennessee	77
Fig. 281.	Bowl, with a handle representing a bird's head and neck, and a projection at the opposite side; Illinois	77
Fig. 282.	Bowl, with four ears set around the shoulder; Kentucky	77
Fig. 283.	Bowl, showing four projections in the plane of the rim; Tennessee	77
Fig. 284.	Bowl, provided with two mutilated studs projecting below the shoulder; Arkansas	77
Fig. 285.	Ornamented Vessel, contracting toward the aperture without forming a shoulder; North Carolina	77
Fig. 286.	Vessel, provided with a low wide neck; Tennessee	77
Fig. 287.	Ornamented flat-bottomed Vessel, with a wide cylindrical neck; Louisiana	77
Fig. 288.	Small-necked, nearly globular Vessel; Tennessee	78
Fig. 289.	Ornamented Vessel, with a wide neck; Georgia	78
Fig. 290.	Flat-bottomed ornamented Vessel, with a narrow low neck; Louisiana	78
Fig. 291.	Long-necked Bottle, ornamented with studs; Tennessee	78
Fig. 292.	Pitcher, with ornamented neck; Utah Territory	78
Fig. 293.	Large Vessel, with a slightly projecting rim and nearly conical bottom; outside marked with impressions probably produced by modelling in a woven basket; Georgia	79
Fig. 294.	Ornamented Vessel, showing a depression around its middle; Louisiana	80
Fig. 295.	Fish-shaped Vessel; Tennessee	80
Fig. 296.	Nearly globular Vessel, with a neck resembling an animal's head, one side of which forms the aperture; Tennessee	80
Fig. 297.	Small globular Vessel, with a neck formed in imitation of a human head; aperture in the occipital portion; Kentucky	80
Fig. 298.	Vase, elaborately ornamented with figures in relief; Mexico	82
Fig. 299.	Pitcher, with two mouths and two handles; highly ornamented with raised figures; Mexico	82
Fig. 300.	Hieroglyphical Tablet	83
Fig. 301.	Vase, showing elaborate raised ornamentation	84
Fig. 302.	Small Vessel, tapering to a point opposite the aperture; Mexico	84
Fig. 303.	Small goblet-shaped Vessel; Mexico	84
Fig. 304.	Human Head; Alabama	84
Fig. 305.	Wolf's Head; Alabama	84
Fig. 306.	Seated Human Figure with another on its back; Mexico	86
Fig. 307.	Seated Human Figure with a peculiar head-dress; Mexico	86
Fig. 308.	Squatting Female Figure; Mexico	86
Fig. 309.	Standing Female Figure with a long gown; Mexico	86
Fig. 310.	Snake, coiled on the back of a Turtle; Mexico	87
Fig. 311.	Coiled Rattle-snake; Mexico	87
Fig. 312.	Ornamented Spindle-whorl; Mexico	87
Fig. 313.	Ornamented Spindle-whorl; Mexico	87

xiv   LIST OF ILLUSTRATIONS.

Objects of Wood.

		PAGE
Fig. 314.	Bailing-vessel; California	88
Fig. 315.	Toy Canoe; California	88
Fig. 316.	Sword-shaped Implement; California	88

Objects of chipped and ground Stone (Supplement).

Fig. 317.	Chipped Perforator, three-sided; flint; California	90
Fig. 318.	Chipped Perforator, massive opposite the point; flint; California	90
Fig. 319.	Chipped Implement, sickle-shaped; flint; Ohio	90
Fig. 320.	Ground Implement, club-head-shaped; greenstone; California	90
Fig. 321.	Ground Implement, curved, and provided with a shoulder on the upper side, and four conical projections on the lower one; basaltic rock; Oregon	90
Fig. 322.	Ground Implement, flat, tapering at one end and broad at the other, where it is pierced; argillite; Massachusetts	90

Hafted Stone Weapons.

Fig. 323.	Grooved Axe; greenstone; Dakota Indians	93
Fig. 324.	Celt; argillite; Indians of the Missouri Valley	93
Fig. 325.	War-club, formed of a round stone firmly attached to a handle; Dakota Indians	93
Fig. 326.	War-club, consisting of a round stone connected with the handle by flexible thongs; Apaches	93
Fig. 327.	War-club with an egg-shaped head of limestone; Blackfeet	93
Fig. 328.	War-club with a head of an elongated egg-shape; greenstone; Missouri River Valley	93
Fig. 329.	Knife with a lance-head-shaped blade of slate; Nunivak Island, Alaska	93
Fig. 330.	Wooden Scabbard for the same	93

Hafted Stone and Bone Tools.

Fig. 331.	Hammer with a head of greenstone; Fort Simpson, British Columbia	95
Fig. 332.	Adze-shaped Pick of whalebone; Mackenzie's River District	95
Fig. 333.	Pick of walrus ivory; Nunivak Island	95
Fig. 334.	Hoe, made of the shoulder-blade of a buffalo; Arickarees, Dakota Territory	95
Fig. 335.	Reaping-hook, made of the lower jaw of an antelope; Caddoes, Indian Territory	95
Fig. 336.	Adze-like Implement, consisting of a small celt-shaped blade of argillite connected with a forked handle; Vancouver's Island	95
Fig. 337.	Celt-like Chisel of argillite, attached to a cylindrical handle; Vancouver's Island	95
Fig. 338.	Celt-shaped Chisel of argillite, connected with a handle of peculiar form; Vancouver's Island	95
Fig. 339.	Flint Scraper, connected with a hook-shaped ornamented handle of elk-horn; Mandans	95
Fig. 340.	Tool of deer-horn, used in chipping stone arrow-heads, perforators, etc.; Nevada Territory	95

INTRODUCTION.

The National Museum has been for years the depository of large and valuable collections illustrative of North American Ethnology, which now form one of its most important departments. In classifying this rich material for the purpose of exhibition during the Centennial Celebration at Philadelphia, it has been thought proper to separate the objects supposed to belong to times anteceding the European occupation of the continent from those that are known to have been manufactured within the period of contact between the Indian and the Caucasian. Only thus it became possible to exhibit, approximately at least, the aboriginal state of culture before it had been modified by European influences. The first or *archæological* series, to which the following account more particularly refers, comprises objects found in mounds and other burial-places of early date, on and below the surface of the ground, in caves, shell-heaps, etc.,—in fact all articles of aboriginal workmanship that cannot with certainty be ascribed to any of the tribes which are either still in existence or have become extinct within historical times. These relics, consisting of chipped and ground stone, of copper, bone, horn, shell-matter, clay, and, to a small extent, of wood, have been grouped according to material, and then classed under such denominations as their forms suggested. Similarity of shape afforded the principal guidance in arranging these specimens, many of which leave a wide scope for conjecture as to the uses to which they were applied by their makers. The second or more strictly *ethnological* series, a description of which is not attempted at present, consists of articles obtained from existing native tribes by private explorations as well, as by expeditions undertaken by order of the United States Government, and contains almost every object tending to illustrate their domestic life, hunting, fishing, games, warfare, navigation, traveling by land—in short every phase of their existence that can be represented by tangible tokens. The use of these objects, many of which show forms copied from the manufactures of the whites, is in most cases well understood, and they have been arranged according to their mode of application, and without reference to the substances of

INTRODUCTION.

which they are made. This mode of classification, as stated, could not be applied to the relics composing what is called the archæological series, considering that the latter embraces a large number of specimens, and even classes of typical objects, to which it would be hazardous to assign a definite use; and this uncertainty attaches even to such common relics of the aborigines as have hitherto been thought to represent well-recognized types. Collectors, for instance, are very ready to class chipped stone articles of certain forms occurring throughout the United States as arrow and lance-heads, without thinking that many of these specimens may have been quite differently employed by the aborigines. Thus the Pai-Utes of Southern Utah use to this day chipped flint blades, identical in shape with those that are usually called arrow and spear-points, as knives, fastening them in short wooden handles by means of a black resinous substance. Quite a number of these hafted flint knives (Fig. 1) have been deposited in the collection of the National Museum by Major J. W. Powell, who obtained them during his sojourn among the Pai-Utes. The writer was informed by Major Powell that these people use their stone knives with great effect, especially in cutting leather. On the other hand, the stone-tipped arrows still made by various Indian tribes are mostly provided with small slender points, generally less than an inch in length, and seldom exceeding an inch and a half, as exemplified by many specimens of modern arrows in the Smithsonian collection. If these facts be deemed conclusive, it would follow that the real Indian arrow-head was comparatively small, and that the larger specimens classed as arrow-heads, and not a few of the so-called spear-points, were originally set in handles and were used as knives and daggers. In many cases, further, it is impossible to determine the real character of small leaf-shaped or triangular objects of chipped flint, which may have served as arrow-heads or either as scrapers or cutting tools, in which the convex or straight base formed the working edge. Certain chipped spear-head-shaped specimens with a sharp straight or slightly convex base may have been cutting implements or chisels. Arrow-heads of a slender elongated form pass over almost imperceptibly into perforators, insomuch that it is often impossible to make a distinction between them. Among the implements, weapons, etc., that have been brought into shape by pecking or grinding there are many types of unmistakable character, such as axes, adzes, mauls, mortars, pestles, pipes, etc.; yet here, too, not a few classes of objects are met to which a definite use cannot be ascribed. Among the latter are disc-shaped stones, pierced tablets, tubes, rings, pendants, and various other

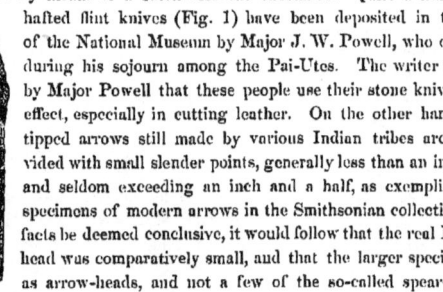

Flint Knife in wooden handle (1).

typical articles. In many instances it cannot be determined whether an object was designed for use or for ornament.

In order to classify the numerous articles composing the archæological series, it was necessary, of course, to arrange them under different heads; but in consideration of their too often doubtful mode of application it cannot be asserted that the specimens represent in all cases the characters attributed to them by the titles under which they have been classed. Nor does the division into two groups intended to illustrate different periods warrant absolute exemption from errors, considering that a number of the articles embraced in the archæological series may have been made after the arrival of the Caucasians in North America, especially such relics as are derived from districts inhabited by tribes that became in comparatively recent times acquainted with the manufactures and commodities of the whites. Yet, after due consideration, the system here adopted seemed better calculated to exhibit the former and present state of the aborigines than any other arrangement presenting the whole available material under one general aspect.

By far the greater number of specimens in the archæological department are manufactures of stone, being fashioned either by flaking or the more tedious process of chipping, or by pecking, grinding and polishing. The chipped series chiefly comprises arrow and spear-heads, cutting and scraping tools, saws, perforators, and digging implements. These articles are usually made of hard silicious stone of conchoidal fracture, such as hornstone, jasper, chalcedony, ferruginous quartz, and other kindred varieties, all of them occasionally comprised in these pages, for the sake of brevity, under the general term "flint," though the real cretaceous flint, which has played such an important part in the prehistoric ages of Europe, does not seem to occur in this country. Many arrow and spear-heads consist of the common white quartz, and some are made of different kinds of stone of inferior hardness. The volcanic obsidian is represented by a beautiful series of Mexican knives and cores, and by arrow-heads, etc., derived from regions north of Mexico. Some Indian tribes still arm their arrows with points of obsidian. In the manufacture of ground and polished weapons, tools and ornaments, the aborigines employed every kind of stone, both hard and soft, suited to their purposes. Grooved axes, celts, adzes, pestles, etc., are very frequently made of varieties of greenstone, a substance which, being hard as well as tough, was well fitted to withstand rough use. Some drilled and highly finished ceremonial weapons are made of the hardest silicious materials, showing that the aborigines were in this respect in advance of the prehistoric races of Europe, who scarcely ever attempted to drill stone of such hardness. Quartzite, sandstone, serpentine, hematite and slate often constitute the materials of ground articles. More precise statements will be made in the proper places.

Though the Smithsonian collections chiefly embrace aboriginal manufactures, ancient and recent, derived from the northern half of the continent, or, in other words, from the vast territory bounded by the Atlantic and the Pacific, the arctic regions and the southern frontier of Mexico, it possesses, in addition, many valuable specimens, and even large collections, from the Antilles and from Central and South America. Perhaps the most important of these collections is one from Porto Rico, presented by the late Mr. George Latimer, for a long time a resident of that island. It comprises many specimens of pottery of a peculiar character, and several hundred articles of stone, among them one hundred and twenty-seven celts, numerous pestles, masks, rubbing-stones, and, above all, a rich series of those curious oval or horse-collar-shaped objects, which have for many years attracted the attention and elicited the comments of archæologists, both in Europe and in America. This collection is probably unsurpassed by any other derived from the Island of Porto Rico. The Central American States are represented by hundreds of specimens of pottery and objects of stone, some of them of remarkable character. The large stone idols obtained by Mr. E. G. Squier in Nicaragua, and described and figured by him in his well-known work on that State, are among the most valued relics of the National Museum. Peru has furnished a large collection of pottery, consisting of one hundred and twenty vessels moulded in the peculiar style formerly prevalent among the aborigines of that country, and also a number of mummies, or rather desiccated human bodies. The other parts of South America—Chile, Guiana, Brazil, and even the southernmost region of the continent, Tierra del Fuego—have likewise contributed their share to enrich the Museum of the capital.

In conclusion, it should be stated that the Smithsonian collections are not derived exclusively from America, but that they likewise embrace manufactures of many races of other parts of the world. Thus, there may be seen in the Museum a great variety of relics pertaining to the prehistoric ages of Europe, such as rude flint implements from the drift of France and England, articles of stone, horn and bone found in the celebrated caves of the Dordogne (Southern France), a large and varied series of Swiss lacustrine antiquities, and many neolithic weapons and tools from Denmark and other districts of Northern Europe. Still more numerous are weapons, utensils, textile and ceramic fabrics from Asia, Africa, Australia, and the island groups of the Pacific. Many of these products of art, including the boomerang of the Australian savage and the carved war-club of the Feegeean, as well as the finished tissues and implements of China and Japan, were procured in the course of explorations undertaken at the expense of the United States Government, as before stated. Among them the circumnavigation of the globe under the command of Captain Wilkes and Perry's expedition to Japan deserve special mention.

The following descriptions refer only to the *typical* objects in the collection. The classification might have been much extended by the introduction of subdivisions, if the character of this publication had permitted a more exhaustive treatment of the subject. The present condensed account is but the forerunner of more minute archæological and ethnological works, which will be published in due time under the auspices of the Smithsonian Institution.

I. STONE.

Archæological researches in Europe have shown that the early inhabitants of that continent used for a very long period exclusively rude tools and weapons of chipped flint, until they began to render their implements of war and peace more serviceable by the process of grinding. Archæologists, therefore, divide the European stone age into a period of chipped and one of ground stone, or, technically speaking, into a *palæolithic* (old-stone) and a *neolithic* (new-stone) period. Palæolithic implements occur in ancient beds of river-gravel and in cave-deposits of early date, and are often associated with the osseous remains of the mammoth, woolly-haired rhinoceros, cave-bear, cave-lion, and other pachydermatous and carnivorous animals now extinct in Europe. The implements of the later or neolithic period indicate a more advanced state of human development, and the animal remains sometimes found with them belong to species still existing in Europe, or known to have there existed within historical times. Thus the gradual progress in the mechanical skill of the prehistoric European is illustrated by his works of art, which present, as it were, an ascending scale, beginning with the rude flint flake or the roughly fashioned hatchet-blade, and terminating with the elaborately chipped dagger or lance-head, the pierced axe, and other types in vogue immediately before the introduction of bronze.

In North America chipped as well as ground stone implements are abundant; yet they occur promiscuously, and thus far cannot be respectively referred to certain epochs in the development of the aborigines of the country, and hence the here adopted separation of North American stone articles into a chipped and a ground series has no chronological significance whatever, but simply refers to the modes of manufacture.

A. FLAKED AND CHIPPED STONE.

1. Raw Material.—As such may be considered pieces of flint, etc., rudely blocked out and presenting no definite form. The Museum possesses a series of these roughly prepared fragments, which were obviously designed to be made into implements. They are often of comparatively large size, and generally consist of some kind of silicious material (hornstone, jasper, etc.). They occur, sometimes many of them together, in various parts of the United States.

2. **Irregular Flakes of Flint, Obsidian, etc., produced by a single blow.**—Some may represent cutting tools of the most primitive kind.

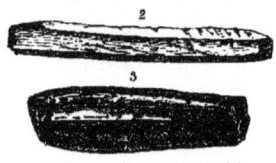

OBSIDIAN KNIFE AND NUCLEUS (½).

3. **Two-edged narrow Flakes of Obsidian and prismatic Cores or Nuclei, from which such Flakes have been detached by pressure** (Figs. 2 and 3, Mexico).—The mode of manufacture of these flakes or knives has been described by some of the early Spanish authors on Mexico.[1] Obsidian breaks like the cretaceous flint of Europe, and hence the Mexican knives are identical in shape with the neolithic flint knives found in the countries bordering on the Baltic Sea.

4. **Pieces of Flint, Quartz, Obsidian, etc., roughly flaked, and either representing rude tools, or designed to be wrought into more regular forms.**—Unfinished Arrow and Spear-heads.

5. **Arrow-heads.**—They are the most abundant aboriginal relics in the United States; but being chiefly made of hard and brittle silicious materials, they were easily damaged in hitting the object at which they were aimed, and many of them consequently bear the marks of violent use. Yet perfect specimens are by no means scarce. The art of arrow-making survives to the present day among certain Indian tribes inhabiting parts of the United States not yet settled by whites, and the National Museum contains a large number of modern stone arrow-heads (partly in shafts) which equal, and even surpass in workmanship, the best specimens picked up in fields or recovered from old Indian burial-places. The modes of their manufacture have been witnessed and described by explorers, and these operations now appear less difficult than they were formerly supposed to be.

A classification of the arrow-heads with regard to their chronological development is not attempted, and hardly deemed necessary. North American Indians of the same tribe (as, for instance, the Pai-Utes of Southern Utah)

---

[1] The fullest account is given by Torquemada (*Monarquia Indiana*, Seville, 1615). The Aztec artisan, he states, dislodged the obsidian flakes from the block by pressure, employing a large wooden T-shaped implement, which acted somewhat in the manner of a punch, the cross-piece resting against the chest. A translation of Torquemada's description is to be found in E. B. Tylor's "Anahuac," London, 1861, p. 331. Motolinia makes similar statements, which, it is believed, have not yet been quoted in English works.

arm their arrows with stone points of different forms, the shape of the arrowhead being with them merely a matter of individual taste or of convenience. It is here only intended to present the characteristic types of these weapons. Yet any such arrangement must be arbitrary to a great extent, owing to the many intermediate forms in which the distinguishing peculiarities are wanting, and the same difficulty is met in the classification of stone articles in general, may they be chipped or ground.

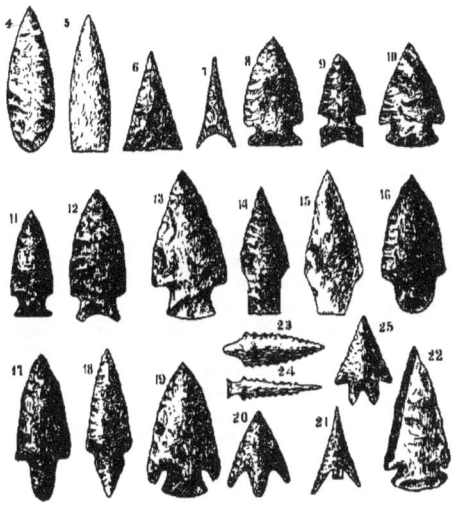

ARROW-HEADS (½).

a. Leaf-shaped, base pointed or rounded (Fig. 4, gray semi-opal, California). Those with a pointed base imperceptibly pass over into the lozenge form, which is not very frequently met.

b. Convex-sided with truncated base (Fig. 5, transparent obsidian, Mexico). Specimens of this description often approach the triangular shape.

c. Triangular, forming an equilateral or isosceles triangle (Fig. 6, gray jasper, New York). Perfectly triangular arrow-points are less frequent than those of the following class.

d. Straight-sided with more or less concave base. In some the concavity assumes the character of a deep indentation by which barbs are produced

(Fig. 7, brown jasper, Oregon). There are varieties of this type, in which the sides appear more or less convex, or straight near the base to a certain distance, where they form obtuse angles or shoulders from which they converge to the point.

e. Notched at the sides near the base, which is straight (Fig. 8, jaspery agate, Texas), concave (Fig. 9, light-brown hornstone, Tennessee), or convex (Fig. 10, gray hornstone, Ohio).[2]

f. Stemmed.—Expanding stem, base straight (Fig. 11, light-colored flint, Ohio), concave (Fig. 12, dark-gray hornstone, Pennsylvania), or convex (Fig. 13, silicified wood, Ohio).—Straight-sided truncated stem; sides of stem parallel (Fig. 14, gray hornstone, Ohio), or converging toward the base (Fig. 15, quartz, District of Columbia). In such specimens the base of the stem is straight or concave.—Rounded or more or less tapering stem (Fig. 16, light-brown flint; Fig. 17, brownish hornstone; Fig. 18, gray hornstone. All from Tennessee). With the arrow-heads characterized by a tapering stem may be classed those of a perfect lozenge form, which, as stated, are comparatively scarce.

g. Barbed and stemmed.—There is much difference in the shape and length of the barbs, and the stems are truncated, rounded or tapering, etc., thus presenting nearly all the forms seen in unbarbed stemmed arrow-heads (Fig. 19, gray-brown hornstone, Tennessee; Fig. 20, brown semi-opal, Oregon; Fig. 21, green semi-opal, Oregon).

In addition, many arrow-heads, belonging by their general shape to one or the other of the classes just enumerated, are modified in different ways. The peculiarity of some consists in their being beveled along both edges on opposite sides, so as to form in the cross-section a figure resembling a long-stretched rhomboid (Fig. 22, gray flint, Tennessee); others exhibit serrated edges (Fig. 23, gray flint, Oregon; Fig. 24, yellow jasper, Louisiana); and in a number of specimens the stem is bifurcated (Fig. 25, gray jasper, Tennessee).

6. **Spear-heads.**—The articles brought under this head are almost as varied in shape as those designated as arrow-heads, and in many instances they present exactly the same forms, the only distinguishing feature being their larger size.[3] As before stated, many of the so-called spear-heads may have been inserted in wooden handles, to serve as cutting tools.

---

[2] In quite a number of notched flint arrow-heads with convex base, and also in many spear-heads (?) of corresponding shape, the curved base-edge exhibits a marked *polish*, as though they had been employed as scraping or smoothing tools. The polish is not intentionally produced, but evidently the result of a long-continued use, totally different from that for which these articles would seem to have been designed.

[3] In separating arrow-heads from the larger objects of similar shape, the writer follows a usage rather than his own inclination.

a. Triangular or more or less convex-sided, sometimes very slender; base straight (Fig. 26, light-gray chalcedony, Tennessee), concave (Fig. 27, yellow jasper, California), or convex, in some cases bluntly pointed.

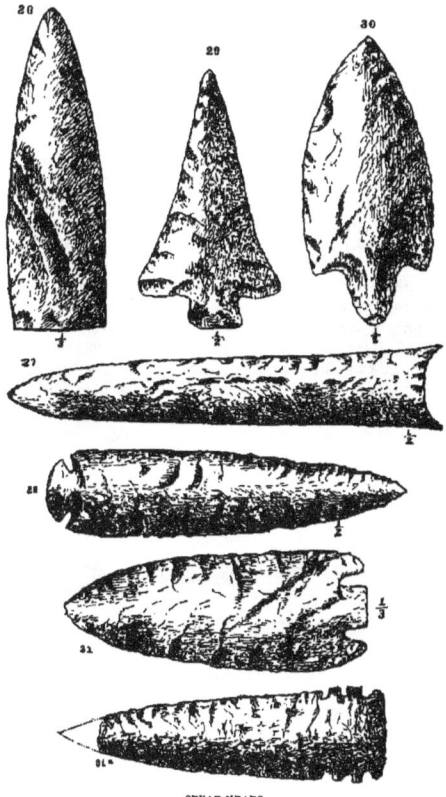

SPEAR HEADS.

b. Notched at the sides near the base, the latter being straight, concave, or convex (Fig. 28, gray flint, Kentucky). Barbs are sometimes formed by the notching, and the beveling on opposite sides, as in arrow-heads, is occasionally to be noticed. Quite exceptional are spear-heads exhibiting several notches at the base (Fig. 31 a, brown jasper, Maine; half size).

c. Stemmed.—Expanding stem, base straight (Fig. 29, quartz schist, Pennsylvania), concave or convex.—Straight-sided truncated stem with parallel or converging sides, and straight, concave, or slightly convex base.—Rounded or more or less tapering stem (Fig. 30, gray flint, New York).

d. Barbed and stemmed (Fig. 31, white milky quartz, Louisiana).

7. Perforators.—The ruder implements of this class may be characterized in a general way as irregular fragments of flint, etc., mostly of an elongated form, which have been chipped to a point at one extremity, and hence it may be imagined that they assume an almost endless variety of shapes. The pointed part, however, presents, from necessity, a more or less developed pyramidal form. Other perforators are worked into shapes sufficiently defined to permit a classification. Yet in many cases it is extremely difficult to distinguish a well-made perforator from a slender arrow-head, especially when the former bears no traces of use at its point. This apparently intact state can be frequently noticed, and hence some persons have gone so far as to deny the existence of North American piercing implements of stone. They forget that the perforating of soft substances, such as moistened hides, would have little effect on a tool of hard material. It is known, moreover, that such implements are still made and used by remote Indian tribes. The more regular perforators may be thus classified:—

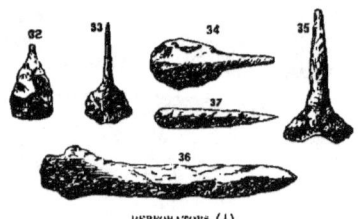

PERFORATORS (½).

a. Almost triangular with broad base and short point (Fig. 32, red jasper, Ohio).

b. Pointed part long and slender, and the opposite end expanding and of irregular outline (Fig. 33, brown jasper, Oregon; Fig. 34, white opaque flint, Missouri).

c. Pointed part long and slender, and expanding base indented, presenting lateral wings (Fig. 35, light-gray flint, Ohio).

d. More or less slender with expanding lower part, which is notched at the sides, or terminates in a stem (Fig. 36, gray hornstone, Tennessee). It may be assumed that perforators of this form as well as of others which afforded no firm grasp to the hand were inserted into handles.

e. Elongated leaf-shape (Fig. 37, gray semi-opal, California).

8. Scrapers.—Thick flakes of flint, obsidian, etc., worked at one extremity into a convex or semi-lunar edge. Some are thus prepared at both ends. These tools were used in cleaning skins, and in scraping and smoothing horn, bone, wood, etc. The Eskimos still use stone scrapers set in well-shaped handles of walrus ivory, horn, or wood. Several specimens of this kind are in the collection of the National Museum.

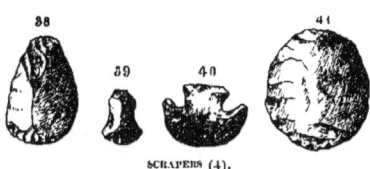

SCRAPERS (¼).

a. Working edge beveled from one side, the lower surface forming a continuous unaltered fracture (Fig. 38, gray flint, Texas). A few are beveled at both ends, and may be called double scrapers. Some terminate in stems opposite the working edge (Fig. 39, compact gray hornstone, Ohio).

b. Working edge chipped from both sides, sometimes at both extremities.

c. Made of the lower portions of broken arrow and spear-heads; working edge chipped from one side or from both (Fig. 40, yellow jasper, Ohio).

d. Disc-shaped, chipped all around (Fig. 41, bluish chalcedony, Texas).

9. Cutting and Sawing Implements.—This group comprises a series of implements which, though differing in form, seem to have been designed for kindred purposes.

a. Flakes of flint and obsidian, more or less chipped at the edges, apparently for the purpose of being used in cutting and sawing (Fig. 42, yellow jasper, Kentucky). The silicious materials out of which such flakes are usually made cannot be split as regularly as the cretaceous flint of Europe, and hence the well-shaped neolithic flakes so frequent in Denmark, Northern Germany, etc., hardly find counterparts among the stone tools occurring north of Mexico. The obsidian flakes from the last-named country, as has been stated, are identical in shape with the corresponding European specimens.

b. Implements with chipped convex edges, mostly serrated at the opposite side, or provided with a row of stems, perhaps for being more securely hafted (Fig. 43, gray flint, California). The specimens of this character were all obtained from California, where the aborigines are known to have employed asphaltum for cementing their stone tools into handles.

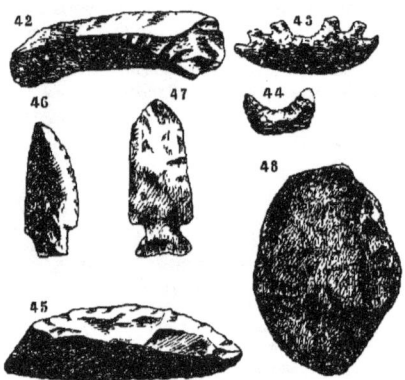

CUTTING TOOLS (½).

c. Small sickle-shaped implements designed, as it seems, for some cutting purpose (Fig. 44, dark-brown jasper, California).

d. Crescent-shaped implements, some of them truncated at one end; probably knives and saws (Fig. 45, lydite, Pennsylvania). A somewhat similar type occurs in Northern Europe.

e. Arrow-head-shaped (notched or stemmed) implements, apparently representing sawing and cutting tools, the part used being either one of the sides which is convex, or the obtuse point (Fig. 46, reddish jasper, Tennessee; Fig. 47, semi-opal, Georgia).

f. Roughly chipped implements with convex edges and massive backs. They resemble the "choppers" found in some caves of Southern France, and described by Lartet and Christy in the "Reliquiæ Aquitanicæ" (Fig. 48, gray hornstone, shell-heap, Maine).

10. Dagger-shaped Implements.—The dagger form is in most cases indicated rather than fully developed. There is, however, in the collection a beautiful specimen remarkable for a well-wrought handle (Fig. 49, gray flint, mound in Alabama). Similar objects are preserved in the Copenhagen Museum.

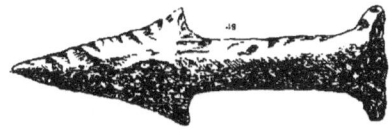

DAGGER (½).

11. **Leaf-shaped Implements.**—Perhaps mostly used for cutting and scraping; some may be unfinished tools.

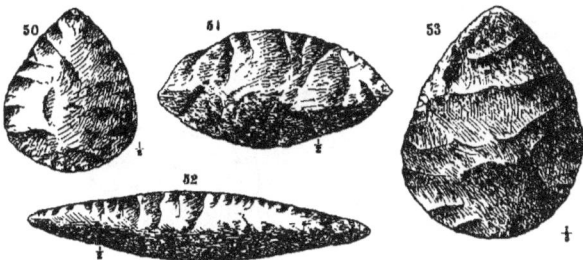

LEAF-SHAPED IMPLEMENTS.

a. Pointed at one end and more or less rounded at the other extremity; sides straight or exhibiting various gradations of convexity. Many specimens of this class present an almond shape, and are thin and sharp-edged. There can be little doubt as to their use as cutting tools (Fig. 50, light-gray flint, Ohio).

b. Approaching an oval shape.

c. Pointed at both ends, broad in the middle, or more or less elongated. They differ much in size, the smaller specimens being not larger than arrow-heads (Fig. 51, brown jasper, Louisiana; Fig. 52, gray flint, Ohio).

d. Large flat implements of roundish, oval, or almond shape, either rudely blocked out, or chipped with more or less care around the circumference. Some appear slightly worn at the edge, as though they had been used for scraping purposes. They occur mostly in mounds and in deposits under the ground, sometimes comprising many hundred specimens. Such deposits have been met from the Mississippi to the Atlantic States. The implements in question frequently consist of the peculiar stone of "Flint Ridge," an elevation extending through Licking and Muskingum Counties in the State of Ohio. The material was here quarried by the aborigines, who have left the traces of their operations in the shape of numerous pits and of accumulations of chips heaped up around them.[4] Many of the specimens closely resemble in shape and size the "hatchets" of the European drift, which occur associated with the remains of extinct animals (Fig. 53 represents a common form. The original belonged to a regular deposit of about fifteen hundred specimens, which was discovered at Beardstown, Cass County, Illinois).

12. **Large flat Implements** of silicious material, usually ovoid in shape, and sharp around the circumference. Some expand considerably at the broader or cutting edge, exhibiting a tapering or truncated opposite extremity (Fig. 54, fine-grained quartzite, Tennessee; Fig. 54 a, same material, Illinois).—The broad part sometimes appears almost glazed from constant wear. They are supposed to have been used as spades or hoes.

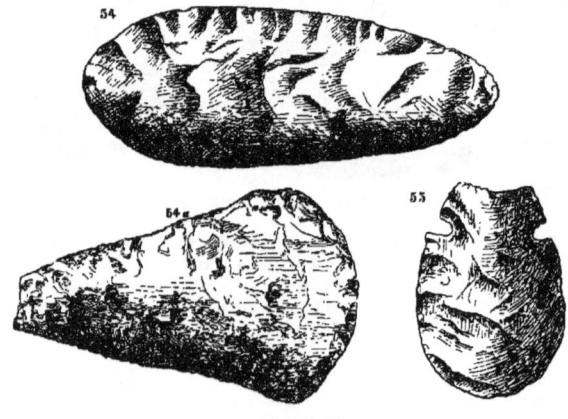

DIGGING TOOLS (⅓).

13. **Large flat Implements** mostly of oval outline, but truncated and laterally notched at the end opposite the working edge (Fig. 55, Illinois).—

---

[4] The locality is described in Squier's "Antiquities of the State of New York," Buffalo, 1851, p. 126.

The lower portion is often smoothed by wear. These implements, like the preceding kind, probably were attached to handles and used in digging the ground for agricultural and other purposes. Both varieties consist of corresponding materials, and sometimes occur together in mounds and subterranean deposits.

14. Wedge or Celt-shaped Implements.—They consist mostly of silicious materials, and bear some resemblance to the rough-hewn flint celts of Northern Europe.

B. PECKED, GROUND AND POLISHED STONE.

1. Wedges or Celts.[5]—They form a numerous class of North American implements, occurring on the surface of the soil and occasionally in mounds, and were doubtless applied to different uses for which their shape and size suited them. They are sometimes rudely pecked or chipped into form, and merely sharpened at the cutting edges; but in general they are entirely ground, and not a few of them exhibit a beautiful polish. Their length varies from an inch and an inch and a half to a foot and more. They consist of different kinds of stone, such as diorite, syenite, hornblende rock, serpentine, etc., and even soft slates have sometimes furnished their material.[6] Occasion-

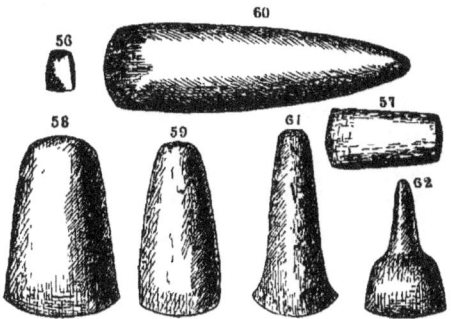

CELTS (¼).

ally specimens made of silicious varieties (hornstone, jasper, lydite) are met, and very small celts consisting of hematite occur in different parts of the

---

[5] From the Latin word *celtis* or *celter*, a chisel.
[6] In Mexico celts of jade are not unfrequent.

United States. They are sharp-edged and highly polished, and were evidently used for cutting purposes (Fig. 56 represents one of these diminutive hematite tools, which was found in Ohio). A cross section parallel with the cutting edge of a North American celt presents in general a roundish or oval outline; but some specimens are four-sided, insomuch that a section would resemble a rectangle with sharp or rounded angles and more or less convex sides (Fig. 57, greenstone, Indiana). The cutting edges, nearly always ground from both sides, are usually convex, and rarely straight. The butt-ends generally exhibit more or less rounded contours (Fig. 58, syenite, Illinois; Fig. 59, greenstone, Tennessee); but in some specimens the butt tapers and terminates in a blunt point (Fig. 60, indurated chlorite slate, Tennessee, mound). Some have expanding cutting edges (Fig. 61, Louisiana). The butts of many celts are much battered, as though the implements had been employed in connection with mallets for splitting wood, etc.; others bear the traces of having been inserted in shafts to serve as axes or adzes. In rare cases the extremity opposite the edge terminates in a sort of a handle (Fig. 62, greenstone, North Carolina). A few specimens of the collection have a cutting edge at each end.

2. **Chisels.**—Wedge-shaped implements of elongated form and comparatively small size have been classed as chisels, and doubtless were used as such. It does not seem that they are abundant. Several specimens of the collection have a round circumference and a greater diameter in the middle or at the blunt end than at the working edge. These implements, which chiefly consist of greenstone, may be considered as typical, having been found in Ohio, Indiana, Virginia, and Connecticut (Fig. 63, diorite, Ohio). Others are four-sided (Fig. 64, lydite, New York), or flat with rounded smaller sides, and a few specimens of yellow or brownish jasper exhibit in part the original chipping, being only superficially ground. They might be taken for Danish or North German productions of the stone age. Some chisels have working edges at both ends. A specimen of the collection marked "ice-chisel" (Fig. 65, basaltic material, Unalaska Island) presents a peculiar shape, terminating in a sort of handle, which is, however, almost too short for being conveniently grasped. There is a possibility that the implement was hafted. (Compare: Nilsson, "Stone Age," Plate VI, Fig. 135).

3. **Gouges.**—They generally consist of materials similar to those of which celts are made; but they occur in the United States far less frequently than the latter, and appear to be chiefly confined to the Atlantic States. It is supposed that they were employed, besides other uses, in the manufacture of wooden canoes and mortars, which the aborigines hollowed out with the assistance of fire. The gouges were well adapted, by their shape, for removing the charred portions of the wood. These implements vary in length from three inches to a foot. In some the concavity is confined to the lower part (Fig.

66, dark hornstone, New York); in others it extends through their whole length (Fig. 67, Pennsylvania). There are implements which, though exhibiting no concavity, somewhat partake of the character of gouges. They can be likened to celts in which the edged portion is plano-convex, so as to produce a hollow cut. They may, in part, have served as adze-heads. Certain

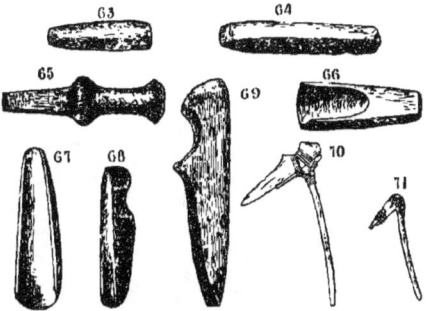

CHISELS, GOUGES AND ADZES (¼).

gouge-like tools (with or without concavities at the cutting edge), which are provided on the convex side with grooves, ridges, or conical elevations, likewise may have formed the heads of adzes, the contrivances just mentioned facilitating their attachment to handles (Fig. 68, greenstone, Massachusetts).

4. Adzes.—There are in the Smithsonian collection some unmistakable adzes—perhaps not very old—derived from the Northwest Coast. One of them (Fig. 69) consists of a dark kind of silicious stone (hornstone), and was obtained in British Columbia. The method of hafting these implements is exemplified by a handled adze (Fig. 70) used by the natives of Oregon. The head, consisting of greenstone, is ten inches long, and connected with the wooden handle by means of split twigs of some flexible kind of wood. There are in the collection other adzes from the Northwest Coast, hafted in a different manner (Fig. 71). In these specimens the small adze-heads of green serpentine are celt-shaped, and rest against a shoulder of the crooked handle, where they are secured by strips of raw-hide, or by cord.

5. Grooved Axes.—Owing to their frequency, these implements may be counted among the best-known relics of the aborigines; and especially in the rural districts of the older States "Indian stone tomahawks" are familiar objects. In general they can be defined as wedges encircled by a groove, usually nearer the butt-end than the edge. The groove served for the reception of a withe of proper length, which was bent around the stone head until both ends met, when they were firmly bound together with ligatures of hide or some other material. The withe thus formed a convenient handle. These axes are frequently made of varieties of greenstone, though specimens consisting of syenite, granite, porphyry, sandstone, etc., are not rare; silicious materials, it seems, were not often employed. Now and then a specimen made of red or brown hematite is met.

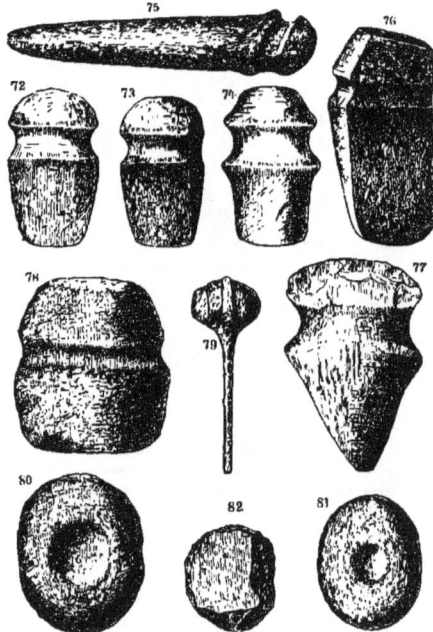

GROOVED AXES, HAMMER-HEADS AND HAMMER-STONES.
(Figs. 72-77: ¼; Figs. 78 and 80-82: ½.)

Grooved axes differ much in size, the smallest in the collection (probably toys) measuring little more than two inches in length and weighing from three to four ounces, while the largest object of this class, a specimen from Illinois (loaned), is thirteen inches long, seven and a half wide, and weighs twenty pounds and a half. Such large tools hardly could be wielded with two hands; yet they must have been employed in some way, their edges exhibiting distinct marks of wear. In general the axes are from five to seven inches long, weighing one and a half or two pounds. In some axes the groove surrounds the stone entirely (Fig. 72, greenstone, Massachusetts), but in others, as it were, only on three sides, the fourth side being flat, and sometimes even slightly hollowed, apparently for resting on a corresponding flat part of the handle (Fig. 73, greenstone, Arizona). The groove is often barely indicated, but deep and regular in the specimens of the better class, which are symmetrically shaped and well smoothed, or even polished. A few specimens exhibit two parallel grooves. The most finished Smithsonian axes, consisting of a dark compact greenstone, are derived from Arizona.

The grooved axes, though corresponding in general form, present many varieties. Their grooves, for instance, are sometimes bounded by ridges, obviously for the purpose of preventing the withe from slipping (Fig. 74, greenstone, South Carolina). In a number of specimens the groove runs obliquely around the stone, which thus evidently formed an acute angle with the handle (Fig. 75, cast, Wisconsin; a specimen of unproportionate length). In rare cases the axes are four-sided, the butt-end terminating in a quadrilateral face (Fig. 76, greenstone, Alaska). In general, however, the butt-ends present rounded contours, and often bear unmistakable traces of violent use. Now and then they are bluntly pointed. The collection contains a few axes with edges at both extremities. Occasionally there occur specimens with remarkably narrow edges (Fig. 77, graywacke, Pennsylvania).

The tools just described are not sharp-edged, and consequently were not used in cutting down trees, but they served for deadening them by the well-known process of "girdling." When the trees had become perfectly dry, they were felled by the application of fire, the axes being again resorted to for removing the charred wood. For the same purpose they may have been employed in the manufacture of wooden canoes. Specimens of small or medium size doubtless were used as battle-axes, like the iron tomahawk of modern times.—No. 7253 of the collection is a cast of the "inscribed" grooved axe found in 1858 on the farm of Samuel R. Gaskill, in Burlington County, New Jersey.

6. **Hammers.**—They comprise hammer-heads and hammer-stones. The former consist of round or oval pebbles, or small boulders of quartzite, granite, greenstone, and other hard and tough materials, and often show no other modification by the hand of man but a groove for the attachment of a handle. Some, however, are artificially brought to the required shape. The groove, it should be stated, is not always carried entirely around the stone. Hammer-heads vary much in size, the smallest specimens measuring only a few inches, while the large ones, designated as mauls, are so bulky and heavy that they could only have been wielded with both hands (Fig. 78, granite, Colorado; eleven pounds). Very large mauls with one or two grooves, sometimes with-

out any groove, have been discovered in the ancient copper mines of the Lake Superior region. They were the tools employed by the aborigines for obtaining the much-valued virgin metal. Some hammer-heads were evidently converted into their present forms from grooved axes whose edges had been damaged by fracture or by constant use. There are in the Smithsonian collection some hafted mauls derived from the Sioux and Assineboins, who still use them for breaking bones, pounding pemmican, etc. (Fig. 79, quartzite, Assineboins; two pounds). These tools, including their handles, are tightly cased in raw-hide, excepting that part of the head which is used for striking. One of these modern handled mauls, derived from the Sioux, is rather heavy, weighing more than nine pounds. The Blackfeet, Sioux, and other still existing tribes sometimes use war-clubs with stone heads. The latter, consisting of quartzite, greenstone, etc., are of a more or less elongated regular egg-shape, well polished, and deeply grooved around the middle for the attachment of the handle. Specimens of this class and of others are in the collection. The different kinds of stone war-clubs in use among the Indians of our time will be described hereafter.

The tools designated as hammer-stones are mostly roundish or oval pebbles of a somewhat compressed or flattened form, presenting in their side view the outline of a more or less elongated ellipse. Quartzite appears to be the prevailing material. Their only artificial alteration consists in two pits or cavities, which form the centres of the opposite broad sides of the pebble. In these cavities the workman placed the thumb and middle finger of the right hand, while the forefinger pressed against the upper circumference of the stone (Fig. 80, quartzite, New York; Fig. 81, quartzite, Pennsylvania). In some instances the depressions are so shallow that they almost escape observation, though specimens with deep and well-defined cavities are not rare. Many hammer-stones bear distinct traces of rough use, being battered and bruised at the circumference. Their longitudinal diameter generally measures from three to five inches, and they may average about a pound in weight.

In Europe similar hammer-stones occur, which have been called *Tilhuggersteene* by Danish archæologists, and it has been conjectured that they were used as tools for chipping weapons and implements of flint. It cannot be doubted that the corresponding American implements served as hammers, since they show the most distinct traces of violent contact with hard substances, and there is much probability that they were used in blocking out flint implements; yet they are by far too clumsy, and possess too much roundness on all sides, to have been the tools for finishing barbed arrow-heads and other delicate articles of flint. Quite different implements were employed in that operation.[7]

---

[7] There are in the National Museum several of the tools employed by modern Indians in the manufacture of stone arrow-heads, perforators, etc. These chipping-implements consist of bluntly pointed rods of deer horn, from eight to sixteen inches in length, or of short slender pieces of the same material bound with sinew to wooden sticks resembling arrow-shafts. The aboriginal "arrow-maker" holds in his left hand the flake of flint or obsidian on which he intends to operate, and presses the point of the tool against its edge, detaching scale after scale, until it assumes the desired form.

There are other quartzite hammer-stones, often of rather irregular shape, in which the cavities are wanting. They have undergone no alteration, excepting that resulting from constant use. A peculiar class of hammer-stones consists of flint pebbles roughly worked into a roundish flattened form. Their battered circumferences indicate the use to which they were applied (Fig. 82, flint, Ohio). Though not in reality belonging to the series of pecked or ground implements, it has been thought proper to mention them in this place. —Certain stones resembling the indented hammer-stones, and often classed with them, evidently were used for other purposes. They will be noticed in connection with mortars.

7. Drilled Ceremonial Weapons.—The grooved tomahawk was among the aborigines, prior to the occupation of the country by Europeans and their descendants, the prevailing implement of the axe kind; but pierced axe and

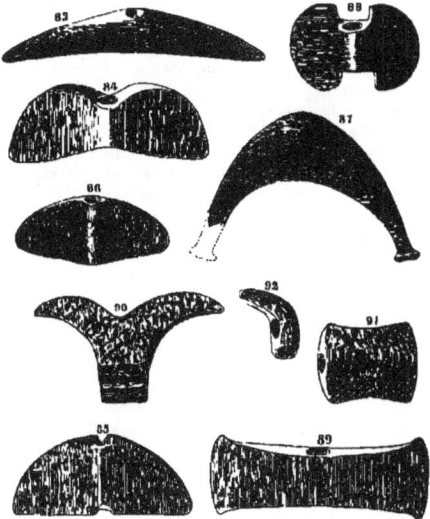

DRILLED CEREMONIAL WEAPONS ($\frac{1}{4}$).

pick-shaped objects also occur, though not in great abundance. These relics are for the most part elegantly and symmetrically shaped, and well polished, but of such small dimensions that they cannot have been applied to any practical use. Their material, moreover, generally consists of soft kinds of stone, more particularly of a gray or greenish slate, which is frequently marked with dark parallel or concentric stripes or bands. Yet specimens made of jasper, ferruginous quartz, syenite, and other hard substances are not wanting. The objects in question doubtless were provided with handles and worn as weapons of parade or insignia of rank by the superiors. They present a great variety of forms, bearing testimony to the ingenuity and good taste of their makers. Many of them somewhat resemble double pick-axes (Fig. 83, serpentine, Virginia; Fig. 84, serpentine, Pennsylvania; Fig. 85, striped slate, Wisconsin; Fig. 86, striped slate, Indiana; Fig. 87, striped slate, Pennsylvania); some are egg-shaped, and others may be likened to axes with two very blunt cutting edges (Fig. 88, cast, original probably brown jasper, Louisiana;* Fig. 89, cast, Wisconsin). In rare cases the parts, which would form the cutting edges in real implements are bifurcated (Fig. 90, striped slate, fragment; Indiana), and in some objects here classed as ceremonial weapons the sides corresponding to edges exhibit a slight inward curve (Fig. 91, translucent ferruginous quartz, Indiana). A few specimens are crooked, terminating in a blunt point at one extremity, and in a rounded butt-end at the other. These specimens are exceptions from the general rule, not being shaped alike on both sides (Fig. 92, striped slate, Indiana).

The holes in these implements have no sufficient width for permitting the insertion of stout handles. They are perfectly regular, and the annular striæ produced by the revolving motion of the drilling tool can often plainly be distinguished. Some specimens, though otherwise finished, are either destitute of shaft-holes, or merely show their beginnings: a fact demonstrating that in North America (as in Europe) articles of this description were first brought to the required shape, and afterward drilled. On the whole, the objects belonging to this class are among the most interesting relics of the aborigines.

8. Cutting Tools.—Any sharpened stone of suitable size could be used as a cutting tool, and hence it may be inferred that the implements of this class assume various forms. Some are of an elongated oval shape, both ends forming cutting edges; others have a crescent shape and vertical cutting edges at both extremities; the most conspicuous form, however, is a flat knife with a semi-lunar edge and a straight back, thick and projecting for greater convenience in handling. These knives chiefly occur in the Eastern States, and their prevailing material is slate (Fig. 93, black slate, Pennsylvania). Yet somewhat similar tools, less defined in shape, but likewise made of slate, were used by the aborigines of the Northwest Coast for ripping open fish. There is in the collection a well-defined cutting tool with a curved edge and a lateral tang,

---

* A beautiful specimen in the collection, exhibiting the shape of Fig. 88, though less elegant in outline, consists of a translucent ferruginous quartz of a pale reddish color. It was found, together with the original of Fig. 91, in Indiana, ten feet below the surface of the ground.

probably serving for the attachment of a handle (Fig. 94, hard red shale, Pennsylvania). Another specimen bearing some resemblance to that just described is provided with a handle of convenient length (Fig. 95, cast, Indiana).

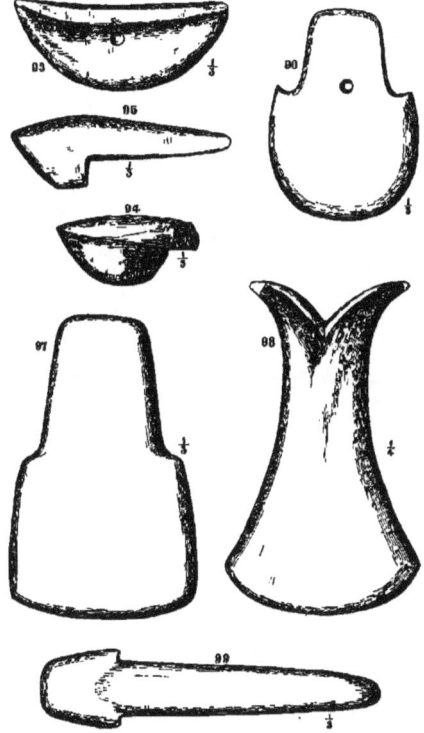

CUTTING TOOLS, SCRAPER AND SPADE LIKE IMPLEMENTS.

9. Scraper and Spade-like Implements.—There is a class of well-finished flattish implements, usually made of compact greenstone, which are formed into a semi-lunar edge on one side, and terminate on the other in a nearly straight-sided handle; a perforation marks the place where the handle and the curved part of the implement meet (Fig. 96, greenstone, Kentucky). These typical objects have been classed as axes, though the smoothness of their edges seems to indicate a different mode of application. It appears more probable that they served as scraping or smoothing tools, and in this case the perforation may have been designed for the reception of a thong, which, passing around the wrist or hand of the operator, enabled him to use the tool with greater force. There are, however, unperforated implements apparently belonging to the same class, in which the handle is almost too broad for convenient use (Fig. 97, east, Arkansas). A cast in the collection deserves particular mention in this place. It is that of a very large tool with a rounded much-used edge, concave sides, and a curious bifurcation at the extremity opposite the working part (Fig. 98, South Carolina). It is not intended to assign any definite use to this remarkable relic. In connection with the tools just mentioned reference may be made to others somewhat resembling diminutive spades, although it is not asserted that they were used as such (Fig. 99, east, South Carolina). These implements seem to be rare. The best specimen known to the writer (represented by a cast in the collection) is in possession of Dr. Joseph Jones, of New Orleans, and was found by him in a Tennessee grave-mound. It consists of greenstone, and measures seventeen inches and a half.[9]

10. Pendants and Sinkers.—The names "pendants" and "plummets" have been given to a class of symmetrically shaped and well-finished objects, which were evidently designed for suspension, though it is not quite certain for what special purpose or purposes they were used. On account of their shape and the pains bestowed on their production they have been classed among aboriginal ornaments; yet the former inhabitants of this country devoted much time and labor to the manufacture of objects of a useful character, and hence it appears not improbable that the articles in question were, in part at least, weights for fishing-lines. These pendants or plummets usually consist of hard materials, such as red or brown hematite, jasper, ferruginous quartz, greenstone, etc. Some are nearly pear-shaped, though more or less elongated, and either entirely smooth (Fig. 100, hornblende rock, Ohio), or grooved near the more tapering end (Fig. 101, red hematite, Tennessee), or pierced with a hole at the same place (Fig. 102, amygdaloid, Arkansas). It is significant that similarly shaped and pierced leaden sinkers for fishing-lines are sold in the hardware stores of this country. Some articles of the class under notice exhibit more developed and really elegant outlines (Fig. 103, greenstone, Ohio). A few specimens, apparently partaking of a kindred character, are of a double conoid form (Fig. 104, greenstone, California). Another of the many varieties expands at the upper end and terminates in a knob (Fig. 105, quartzite, Massachusetts).

---

[9] Figured in "Antiquities of the Southern Indians," by Charles C. Jones, Plate XVII, Fig. 2.

Specimens worked with less care are not wanting, and among them may be mentioned a variety of an irregular roundish or oval shape, and characterized by a knob at the upper end (Fig. 106, greenstone, Massachusetts). There is

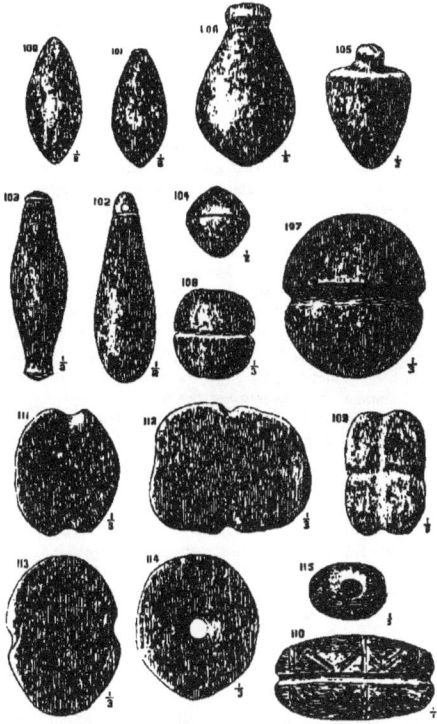

PENDANTS AND SINKERS.

much probability that they served for sinking nets. Some Smithsonian specimens of this description are half a foot long and weigh more than three pounds. The character of net-sinkers appears more distinct in the types following next, to which, indeed, that use has been ascribed by common consent, based upon the fact that net-weights of corresponding shapes are still employed by primitive races of man. Some are roundish stones of various sizes, either worked or left in their natural state, and grooved around the middle for fastening the strings or thongs by means of which they were connected with the nets (Fig. 107, granite, Rhode Island; Fig. 108, potstone, Georgia). It is not always easy to distinguish specimens of this description from grooved hammer-heads. Occasionally a sink-stone exhibits two grooves which cross each other at right angles (Fig. 109, talcose slate, Rhode Island). A small sinker-like specimen of the collection is decorated with engraved lines (Fig. 110, sandstone, Oregon). It may not have been a sinker, but an ornament or an amulet.

A more simple kind of net-sinkers consists of flattish pebbles of roundish or angular (generally indefinite) shape, and of various sizes, which exhibit on two opposite sides of the circumference an indentation or notch, more or less deep, and produced by blows (Fig. 111, quartzite, Pennsylvania; Fig. 112, graywacke, New York; Fig. 113, quartzite, Pennsylvania).[10] In conclusion, the perforated net-sinkers must be mentioned. They are generally made of flat stones of a roundish outline, and exhibit in or near the centre a rather large perforation, which is drilled from both sides in most cases (Fig. 114, micaceous slate, California). These net-sinkers are often made of potstone, as, for instance, in Georgia, where they mark, as elsewhere, the sites of former fishing stations of the Indians. It is not safe, however, to ascribe indiscriminately the character of net-weights to all these pierced flat stones, considering that many of them may have been otherwise utilized.

Much rarer than the sinkers just mentioned are others consisting of pebbles perforated with an oblique hole, not in the centre, but nearer the circumference of the stone. The hole is drilled from two sides, and generally forms an obtuse angle where the perforations meet (Fig. 115, sandstone, Ohio).

11. Discoidal Stones and Implements of Kindred Shape.

The articles enumerated under this head, notwithstanding their resemblance in general form, probably served for different purposes; but what these purposes were, is not always apparent, and the difficulty of classifying the objects in question is enhanced by the almost imperceptible transition from one form into another. Adair, Du Pratz, Lawson, and other early writers have described an Indian game, in which many of the so-called discoidal stones may have been employed. That game is likewise referred to by Lewis and Clarke, Catlin, Murray, and other travelers of more recent times. Speaking of the games in vogue among the Cherokees, Adair describes that diversion in the following words:

"The warriors have another favorite game called *Chungke*, which, with propriety of language, may be called 'Running hard labor.' They have near

---
[10] The writer has seen specimens with four and more indentations.

their state-house a square piece of ground well cleaned, and fine sand is carefully strewed over it, when requisite, to promote a swifter motion to what they throw along the surface. Only one or two on a side play at this ancient game. They have a stone about two fingers broad at the edge, and two spans round; each party has a pole of about eight feet long, smooth and tapering at each end, the points flat. They set off abreast of each other at six yards from the end of the play-ground; then one of them hurls the stone on its edge, in as direct a line as he can, a considerable distance toward the middle of the other end of the square; when they have ran a few yards, each darts his pole anointed with bear's oil, with a proper force, as near as he can guess in proportion to the motion of the stone, that the end may lie close to the stone; when this is the case, the person counts two of the game, and, in proportion to the nearness of the poles to the mark, one is counted, unless by measuring, both are found to be at an equal distance from the stone. In this manner the players will keep running most part of the day, at half speed, under the violent heat of the sun, staking their silver ornaments, their nose, finger, and ear-rings; their breast, arm, and wrist-plates, and even all their wearing apparel, except that which barely covers their middle. All the American Indians are much addicted to this game, which to us appears to be a task

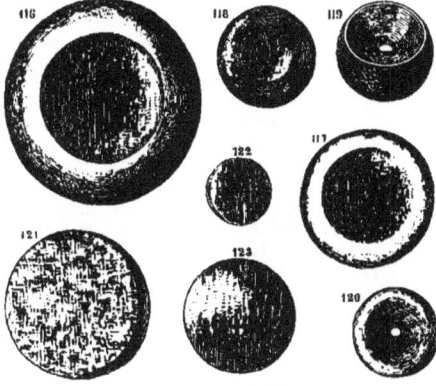

DISCOIDAL STONES (⅓).

of stupid drudgery; it seems, however, to be of early origin, when their forefathers used diversions as simple as their manners. The hurling-stones they use at present were, time immemorial, rubbed smooth on the rocks, and with prodigious labor; they are kept with the strictest religious care from one generation to another, and are exempted from being buried with the dead. They belong to the town where they are used, and are carefully preserved."[11]

There are several kinds of discoidal stones which may have served in the Chung-kee game. Some are quite large, measuring six inches and more in diameter, and bearing a very regular dish-shaped cavity on each side. Their material is often a beautiful (sometimes translucent) ferruginous quartz, and specimens made of this mineral appear to be more numerous in Tennessee than in other States of the Union. The roundness and general regularity of many objects of this class hardly can be surpassed, and not few of them are beautifully polished. In some the outer circumference appears more or less convex, though straight-sided specimens are not wanting (Fig. 116, yellow-brown ferruginous quartz, Tennessee; Fig. 117, brown ferruginous quartz, Tennessee; Fig. 118, dark greenstone, mound in Illinois). In a number of the stones, supposed to have been used in the Chung-kee game, the cavities on both sides are carried somewhat deeper than in the preceding kind, and their centre is marked by a perforation (Fig. 119, east, Ohio; Fig. 120, quartzite, Ohio). These central holes sometimes attain a comparatively large size, imparting to the objects a ring-like character, in which cases it is impossible to state, with any plausibility, whether the specimens, which are, moreover, often somewhat rudely shaped, served as Chung-kee stones, as net-sinkers, or for other purposes.

Some stones, supposed to have been used in the Indian game, show flat or slightly convex circular faces, and perpendicular or even oblique circumferences (Fig. 121, quartzose stone, Georgia).[12] Stones of this description have been called "weights," on account of their resemblance to the iron weights in common use. There are in the collection similarly shaped stone discs of small size, in some cases measuring hardly more than an inch in diameter. Though too diminutive to have served in the Chung-kee game as practised by adults, it is not improbable that children employed them for the same purpose, if, indeed, they were not designed for an altogether different kind of game (Fig. 122, argillaceous material, Pennsylvania).[13] In some instances the discoidal stones assume a lenticular shape, the periphery being represented by a rounded edge (Fig. 123, ferruginous quartz, Texas).

The hollowed discs before described have now and then been taken for mortars in which paint or other substances were pulverized, and the appearance of the concavities in a few lends some probability to that supposition. In those cases, however, they were made to serve a secondary purpose. Specimens with convex or flat faces, again, probably were often utilized as mealing-stones, or for grinding other substances, and some of them may have originally been fashioned for such ends.

The discoidal stones of the perforated kind pass over by slow degrees into

---

[11] Adair: History of the American Indians, London, 1775, p. 401.
[12] See Du Pratz: Histoire de la Louisiane, Paris, 1758, Vol. III, p. 2.
[13] Somewhat similar discs, made of broken clay vessels, are often found on the sites of Indian settlements.

the ring-form, a type exemplified by a large number of specimens obtained from the Californian islands forming the Santa Barbara group. These rings, composed of sandstone, serpentine, potstone, etc., vary much in size and character of workmanship. Some are not more than an inch and a half in diameter, others measure as much as five inches. There are flat specimens not

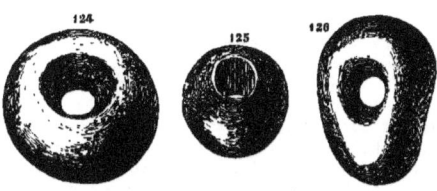

CLUB-HEAD-SHAPED STONES ($\frac{1}{4}$).

exceeding half an inch in thickness, while others are massive, presenting a more or less compressed globular form. There is also much difference in the width of the perforations, which are, however, smooth and round in most cases, though exceptionally of an oval shape. The great variety of forms exhibited in these perforated objects defies all attempts to assign to them anything like a definite use. The more bulky specimens somewhat bear the character of club-heads, and may have been employed as such.[14] Some are of a spherical or conoidal shape, and in the latter the perforation is drilled in the direction of the longer axis. In many the prominent part of the periphery bears the marks of rough use (Fig. 124, hornblende rock, Santa Catalina Island, California; Fig. 125, greenstone, Santa Rosa Island, California). A few of these specimens are of a flattened pear shape, the perforation running in the direction of the shorter axis (Fig. 126, serpentine, Santa Rosa Island).[15] The writer is not aware of the occurrence of such relics in the eastern or middle portions of the United States.

The collection in the Smithsonian Institution contains a series of globular and egg-shaped stones (mostly natural formations) of suitable size to represent

---

[14] It has also been suggested that they served as weights for digging-sticks.

[15] Through the agency of Mr. Paul Schumacher the National Museum has been enriched with a large number of valuable relics from the Californian islands of *San Miguel, Santa Cruz, San Nicolas,* and *Santa Catalina,* and from various points on the main-land, embraced in the Counties of San Luis Obispo and Santa Barbara. A place called *Dos Pueblos* in the last-named district has furnished many remarkable objects. The relics occurred in graves and on the surface. Many are evidently very old; others exhibit a more recent appearance, and some of these have been found in graves with articles of European manufacture (iron knives, objects of brass, beads of glass and enamel, etc.), proving that they are referable to the aborigines whom the whites found in possession of those islands and the neighboring coast. It has been thought proper to include these products of Indian art in the archæological series. The islands have been totally vacated by the Indians, the last of whom, ten in number, were removed, about forty years ago, to the Santa Barbara mission on *terra firma.* A few only are now and then seen in the neighborhood. Mr. H. H. Bancroft mentions in his work, entitled "The Native Races of the Pacific States," the names of some of the tribes formerly inhabiting the localities in question (Vol. I, p. 459, etc.). The graves of *Dos Pueblos,* it should be stated, were also explored by Dr. H. C. Yarrow.

club-heads, and the manner in which some of them, perhaps, were utilized, is illustrated by a number of weapons obtained from existing tribes. There is, for instance, a Sioux war-club with a round stone head about three inches in diameter, and a wooden handle nearly two feet long, the stone as well as the handle being enclosed in a tightly fitting covering of raw-hide sewed together with strong sinew. A loop at the end of the handle serves for attaching the weapon to the wrist. Another kind of stone war-club, represented by a number of specimens in the collection, is still in use among the Apaches, Shoshonees, and other tribes. It consists of a skin-covered stone ball, from two to nearly three inches in diameter, and connected by short thongs with a wooden handle, from eight to twelve inches in length, likewise covered with leather, and provided with a loop at the lower end. The rawhide casing of these weapons, which resemble the "morning-stars" seen in European collections of mediæval armor, consists of one piece, taken from the caudal portion of a bovine. The handle is encased in the close-fitting skin of the animal's tail, a dangling tuft of its hair occasionally forming an ornamental appendage to the weapon.[16]

It may not be amiss to mention in this place certain stones of quartzite, etc., worked into a regular egg-shape, from two to three inches in longitudinal diameter, and slightly truncated at the more pointed end, so as to allow the stone to stand upright on its base. They may have been employed as club-heads, though it appears just as probable that they were used in some game, or perhaps as targets to be shot at with arrows for the sake of practice. Placed upright on a pole, they would fall down when touched by a missile. The specimens in the collection are all derived from Georgia.[17]

12. **Pierced Tablets and Boat-shaped Articles.**—A rather numerous class of aboriginal relics consists of variously shaped tablets of great regularity and careful finish, pierced with one, two, or more round holes. They are mostly made of slate, and the greenish striped variety before mentioned seems to have been preferred by the makers. A very common form is that of a rectangle, with sides exhibiting a slight outward curve. Other tablets are lozenge-shaped with inwardly curved sides, oval, cruciform, etc. Most of them have two perforations, though specimens with only one are not rare, while those that have more than two holes are of less frequent occurrence. The holes are drilled either from one side or from both, and, accordingly, of conical or bi-conical shape. They seldom have more than one-eighth of an inch in diameter. In some tablets the edges are marked with notches, which may be either ornamental, or designed for enumeration. (Fig. 127, slate, New York; Fig. 128, slate, Pennsylvania; Fig. 129, east, Louisiana; Fig. 130, slate, Tennessee; Fig. 131, slate, Tennessee; Fig. 132, striped slate,

---

[16] The clubs here mentioned will be figured hereafter.
[17] These egg-shaped stones have been noticed in the "Antiquities of the Southern Indians" by Charles C. Jones.

Tennessee). Concerning the destination of the tablets nothing is definitely known. At first sight, one might be inclined to consider them as objects of ornament, or as badges of distinction; but this view is not corroborated by the appearance of the perforations, which exhibit no trace of that peculiar

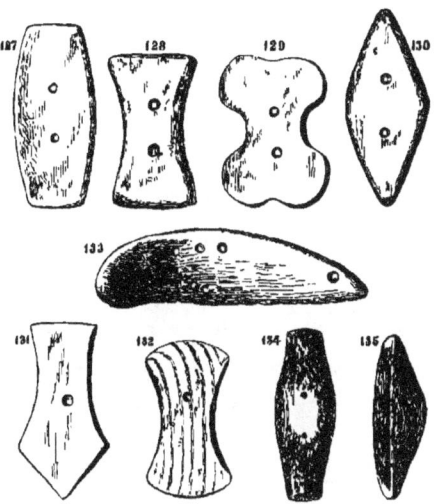

PIERCED TABLETS AND BOAT-SHAPED ARTICLES ($\frac{1}{4}$).

abrasion produced by constant suspension. The classification of the tablets as "gorgets," therefore, appears to be erroneous. There are, indeed, perforated tablets which unquestionably were worn as ornaments; but they will be considered hereafter. Schoolcraft regards the objects under notice as implements for twine-making. According to another conjecture they were used in condensing and rounding bow-strings, by drawing the wet strips of hide, or the sinews employed for that purpose, through the perforations. It is suggestive that the Indians of Southeastern Nevada have been seen using similar pierced tablets for giving uniform size to their bow-strings.[18] There are in the collection some flattened stones of less symmetrical outline, pierced with a number of holes which are rather irregularly distributed, but equal in size to those observed in the tablets just described (Fig. 133, potstone, Penn-

---

[18] Smithsonian Report for 1870, p. 404.

sylvania). Like many other aboriginal relics, pierced tablets occur in sepulchral mounds as well as on the surface of the ground. Those taken from mounds are said to have mostly been found by the side of the skeleton, or near the bones of the hand.

Allied to the pierced tablets are certain boat-shaped articles, either solid or hollowed on one side, and perforated with two holes, mostly of conical form, and placed near the middle or the extremities of the objects. These relics, though agreeing in general character, differ much in the details of their execution, some being of nearly oval, others of rectangular outline, while the cavity, when it occurs, is sometimes shallow, but in other cases so deep as to give the object almost the appearance of a shell. In a few instances the perforations are altogether wanting. Such specimens, however, may have remained in an unfinished state. The objects in question are nearly always well fashioned and polished, their material consisting sometimes of porphyritic syenite, greenstone, etc., but occasionally of softer substances, such as slates, among which the striped variety seems to prevail. Their purpose, probably, was similar to that for which the pierced tablets were designed (Fig. 134, striped slate, Ohio; Fig. 135, greenstone, Kentucky).

13. Stones used in Grinding and Polishing.—There are in the archæological department of the National Museum many stones marked with hollow

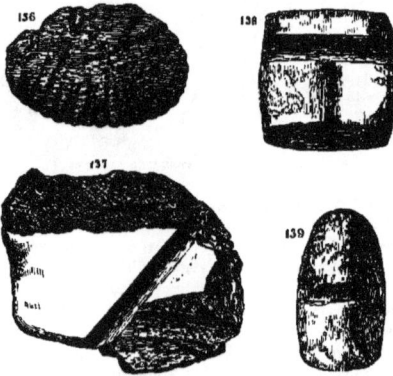

STONES USED IN GRINDING, ETC. (¼).

faces, grooves, or notches, which were apparently produced by the grinding or sharpening of tools, or by the process of smoothing and condensing cords of animal or vegetable material. The more special uses of these relics are not quite obvious in many cases. Certain flattish stones which are furrowed with grooves radiating toward the circumference, may have been used in the preparation of cords (Fig. 136, quartzose rock, New Jersey). On other stones are seen straight grooves of suitable size for straightening and rounding the shafts of arrows (Fig. 137, chlorite slate, Massachusetts; Fig. 138, compact chlorite, Mexico; Fig. 139, hornblende rock, Southern Utah; probably recent). The most conspicuous specimen of this class is a heavy limestone block, bearing on its surface seven deep straight grooves from eight to ten inches in length. This specimen was found in Onondaga County, New York.

In lieu of the grooved stones some Indian tribes of our time employ for fashioning their arrow-shafts short wooden sticks hollowed longitudinally and coated on the inner side with a cement of coarse quartz sand and glue. This aboriginal contrivance is illustrated in the collection by several specimens obtained from the nearly extinct Mandan tribe.

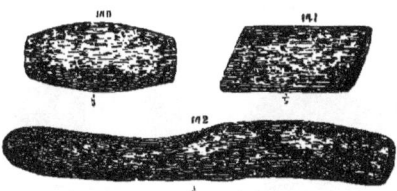

STONES USED IN POLISHING.

No group of aboriginal relics, perhaps, is more difficult to describe than the stones which have apparently served for polishing implements or parts of implements of stone, horn, bone, etc., and, probably, for smoothing leather and other soft substances. In many cases it is by no means improbable that stones supposed to have been used in those operations were otherwise employed. The difficulty of classing these tools is greatly enhanced by the totally unfixed character of their appearance, for nearly every stone of suitable size and furnished with a smooth surface could be utilized as a polisher. There is, for instance, in the collection a piece of yellowish jasper, about three inches and a half in diameter, which exhibits eight perfectly smooth and even facets, each of which presents a different form. It is difficult to assign to this stone any other use than that of a polisher. The collection contains several specimens of similar, though less striking, character. Other polishers are regularly shaped and carefully worked, and nothing indicates their application as polishing tools but the smoothness of those parts with which the operation was performed. One specimen presents the outline of an oval with truncated ends, which, to judge from their glossy appearance, were used in the polishing

36   PECKED, GROUND AND POLISHED STONE.

process (Fig. 140, quartzose rock, Indiana). There is a cast in the collection, presenting the fac-simile of a flat implement of rhomboidal outline, showing very glossy side-surfaces which seem to have been used in polishing (Fig. 141, Louisiana). Other specimens are shaped like very flat celts of equal thickness, in which, as it appears, the blunt edges formed the working parts. It is possible, however, that specimens of this form were intended for other operations. A curious class of implements supposed to have served as polishers, consists of stick or club-shaped stones—mostly natural formations, but sometimes modified by art—which bear at their ends the marks of friction (Fig. 142, lydite, Pennsylvania).

14. Stone Vessels.—Though nearly all classes of aboriginal relics are represented on a large scale in the National Museum, the series of vessels of stone is particularly distinguished by the number as well as by the diversity of the specimens. The most elaborate objects of this kind are derived from the Californian islands (San Miguel, Santa Cruz, Santa Catalina, etc.), and from the opposite coast, a region where the aborigines excelled in various kinds of manufactures.

STONE VESSELS (¼).

It appears that vessels consisting of hard kinds of stone occur rarely in that part of the United States which lies east of the Rocky Mountains. In the Atlantic and Middle States, however, vessels made of the comparatively soft potstone (commonly called soapstone—the *lapis ollaris* of the ancients) have often been met. They differ, of course, in shape and workmanship, some being rather uncouth specimens of aboriginal art; others, again, are tolerably well formed, and betoken no small degree of perseverance on the part of their makers. Most of those seen by the writer were of an elongated shape, somewhat like a boat or a trough, and provided with projections or handles at the opposite narrower extremities (Fig. 143, Massachusetts). A bowl-shaped vessel from Wyoming Territory (Fig. 144) is made of the same material. By far the best potstone vessels, however, have been found in the Californian districts before mentioned.[19] Among them are nearly globular cooking vessels with rather narrow apertures encircled by raised rims. Some of them measure more than a foot in height and fifteen inches in diameter, and their thickness, about five-eighths of an inch at the rim, gradually increases toward the bottom. These utensils are admirable specimens of Indian skill, being almost as regular in outline as though they had been produced with the assistance of the turner's wheel (Fig. 145, Dos Pueblos, Santa Barbara County). Other Californian potstone vessels of large size present the shape of high bowls. One of them is pierced with two small holes near the rim, evidently for repairing the damage produced by a crack (Fig. 146, Dos Pueblos). Among the smaller vessels made of the same material, and obtained from the same region, may be mentioned one which is formed in the shape of a boat (Fig. 147, Santa Cruz Island). Serpentine was likewise employed by the Californian aborigines as the material for vessels, such as cups, bowls, etc., which are in no way inferior to those made of potstone, and even surpass them by being well polished (Fig. 148, serpentine, San Miguel Island). It seems, however, that only small or medium-sized objects of this class were made of serpentine. A small Californian sandstone vessel with an oval aperture, and deeply hollowed, probably served as a drinking cup (Fig. 149, Santa Cruz Island).

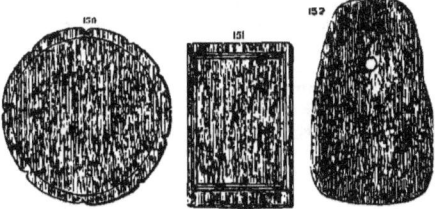

STONE PLATES (⅛).

It may not be altogether out of place to mention in connection with stone vessels a class of remarkable stone plates, which possibly may have pertained to the culinary utensils of the aborigines. One of the specimens is a perfectly

---

[19] These stone vessels as well as the Californian mortars and pestles described on the following pages were recovered from graves by Mr. Paul Schumacher.

flat and well-smoothed stone plate of circular shape, three-eighths of an inch in thickness, and measuring a little less than ten inches in diameter. An incised line runs parallel with the circumference, which is further ornamented with nine rather irregularly distributed notches (Fig. 150, graywacke, mound in Alabama). Another specimen of the same character (derived from the same locality) measures only eight inches in diameter, and is ornamented with three engraved parallel rings and twenty-one notches around the periphery. A third elaborately finished stone plate is of a rectangular shape, and bears as ornaments incised lines which run parallel with the sides, forming three rectangles, and six notches on each of the smaller sides (Fig. 151, material and locality the same). It would be impossible, of course, to state the exact use of such plates, and it remains undecided whether they served as griddles, or as plates for holding solid food, or for some ceremonial or other purpose. A roughly worked plate of clay-slate, nearly rectangular in outline, and measuring about seven inches by five, was found in an Indian grave in Tennessee near the skull (No. 16799 of the collection). This plate and the more elaborate specimens just described possibly were designed for the same use.

There are further to be mentioned slightly concave perforated plates of different sizes and shapes, with angles rounded by the action of the elements rather than by art. They consist of potstone, and were obtained from California (Fig. 152, Santa Cruz Island). The character of the curvature in these Californian plates seems to indicate that they were made from broken vessels. An explanation of their special use is not attempted for the present.

15. Mortars.—The mortars and mortar-like utensils form a particularly rich and varied series in the National Museum, embracing all forms and sizes, from the diminutive cup-shaped stone with a cavity not large enough to hold a hazelnut, and apparently used for grinding pigments, to the ponderous deeply hollowed vessel designed to withstand the operation of the heavy stone pestle. The cultivation of maize among the aboriginal tribes spread over the eastern area of the present United States necessitated the application of grinding utensils, which are, therefore, not unfrequently found on the sites of their former settlements. They are stone slabs or boulders exhibiting shallow concavities, or real mortars hollowed to a depth sufficient to hold a quantity of the cereal. It is shown, however, by the occurrence of circular cavities in projecting ledges of rocks, or in large immovable boulders, that the aborigines sometimes dispensed with portable mortars. Such stationary contrivances for triturating grain have been noticed in many localities where the Indians formerly dwelt. They used also large wooden mortars hollowed with the assistance of fire, as described by Adair in his "History of the American Indians" (p. 416). Some wooden mortars, made by the Iroquois, may be seen in the ethnological department of the National Museum. They are cylindrical, twenty-six inches high, and a little more than fifteen in diameter. The rounded cavity has a depth of about one foot. The wooden pestles used in connection with them measure more than four feet in length.[20] A wooden Mohave mortar of the collection is not quite so large, and not cylindrical, but somewhat tapering toward the bottom. In this specimen the hollowing by fire is distinctly perceivable.

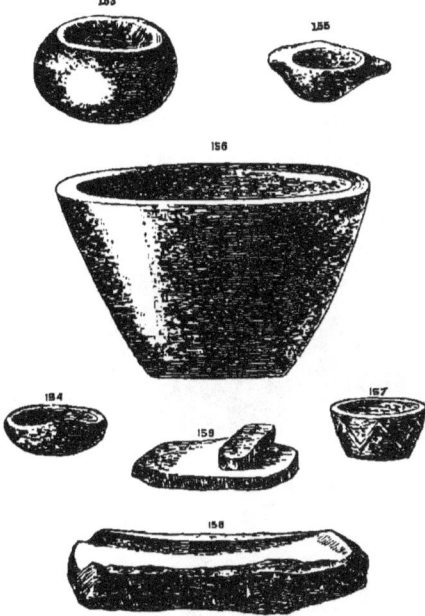

MORTARS AND KINDRED UTENSILS (⅕).

By far the best stone mortars in the Museum have been obtained from the Californian islands and the neighboring coast, more especially from Dos Pueblos. They are made of a compact sandstone which, though of sufficient hardness could be worked with tolerable ease. Some of these mortars are mere boulders hollowed to the proper depth (Fig. 153, San Nicolas Island; Fig. 154, same locality); others have been modified to a certain extent (Fig. 155, same

---

[20] A drawing of an Iroquois mortar with pestle is given in Morgan's "League of the Iroquois," Rochester, 1851, p. 371.

locality). Not a few of them, however, are of a remarkably symmetrical shape, and their production, notwithstanding the tractable character of the material, must have been the result of long-continued patient labor. Many measure more than a foot in height, and nearly twenty inches in diameter at the widest part. They are about an inch and three-fourths thick at the rim, but increase slightly in thickness toward the bottom. The very regular cavity in these mortars reaches a depth of nine and a half inches (Fig. 156, Dos Pueblos). In a number of the mortars the flat rim was inlaid with small pieces of shell, some of which are still in place. They were cemented into the stone by means of asphaltum. A mortar of rather small size, but shaped like the larger specimens, exhibits on its outer side a raised zigzag ornamentation (Fig. 157, Santa Cruz Island).

The mortars thus far described were used in connection with pestles, or, perhaps, sometimes with rounded stones fitting in their cavities, and thus forming crushing tools rather than pounders. Other utensils of a somewhat kindred character are trough-shaped, and the grinding operation was performed by pressing a stone of suitable form forward and backward in the elongated cavity. Several specimens of this description are in the collection. They were obtained (chiefly through the agency of Major J. W. Powell) from Utah Territory, where such utensils, which resemble in general character the Mexican *metate*, are still used by the aborigines (Fig. 158, sandstone). Instead of the concave stone a perfectly even stone slab is employed, in connection with a rubbing-stone with flat faces, by New Mexican tribes (Fig. 159, granite slab, sandstone rubber, Navajo Indians).

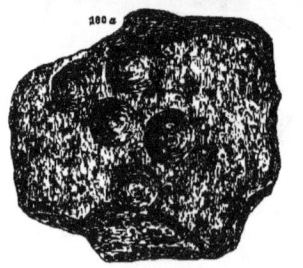

STONES BEARING CUP-SHAPED DEPRESSIONS (⅓)

Somewhat partaking of the character of mortars are good-sized stones, mostly solid slabs, exhibiting on one of the faces, or on both, rather irregular cup-shaped depressions, usually placed near each other. It is supposed that the natives used these stones for cracking nuts which they laid in the cavities, applying a stone for breaking them. There are several of these "nut-stones" in the collection. Specimens made of potstone have been found in districts of Georgia where walnut-trees abound.[11] The Indians, it is well known, made oil from the fruits of these trees (Fig. 160, sandstone, Pennsylvania). There are, however, stones showing (on one side only) artificial cup-shaped depressions of such regularity and smoothness that another use must be ascribed to them. The specimens of the latter class which have thus far fallen under the writer's notice were obtained in Ohio and Kentucky, and their material was sandstone. It is not known whether they were employed, as has been suggested, in some game, or served as receptacles in which paint was rubbed, or for some other purpose (Fig. 160a, sandstone, Kentucky). Not a few of the stones with a cavity on each side, and commonly classed with the hammer-stones previously described, may have served as nut-stones, and others evidently were paint-mortars. Some specimens of the collection still bear the traces of red paint in their cavities.

16. Pestles.— These implements mostly form supplementary parts of mortars, and therefore naturally follow immediately after them in the present enumeration. The specimens in the collection of the National Museum, which can be counted by the hundred, were chiefly derived from the Eastern States, from California and the Northwestern districts. In addition, many have been obtained from other parts of North America. There is considerable difference in their appearance, but the prevailing form seems to be that of a bluntly pointed cone, swelling gradually toward the working portion. Four-sided pestles are of rather rare occurrence. In length pestles vary from a few inches to two feet and more, and their thickness differs accordingly, though not always in proportion, short specimens being sometimes thick and clumsy, while those of considerable length are of a relatively slender and tapering form. Many specimens of the collection were found with the remarkable stone vessels and mortars on the islands of the Santa Barbara group and the opposite main-land. They are partly of the simple conical shape to which allusion was made (Fig. 161, syenite, Santa Cruz Island). This elementary form occurs in many parts of the United States. Other specimens expand at the upper end into a kind of knob (Figs. 162 and 163, compact sandstone, Dos Pueblos). In a third class an annular ridge surrounds the tool below the upper end, which tapers to a blunt point (Fig. 164, sandstone, Dos Pueblos). And lastly, a variety has to be mentioned which exhibits a knob-like expanse at the lower extremity (Fig. 165, amygdaloid, mound at Crescent City, California). In the New England States pestles, more or less resembling a cylinder with rounded ends, are quite frequent, and sometimes of considerable length (Fig. 166, fine-grained sandstone, Rhode Island). Though the extremities of these

---

[11] C. C. Jones, "Antiquities of the Southern Indians," New York, 1873, p. 315.

cylindrical implements bear often the unmistakable marks of wear, it appears probable that they were sometimes used like rolling-pins for crushing the grain. A very fine specimen from Alaska, measuring as much as two feet

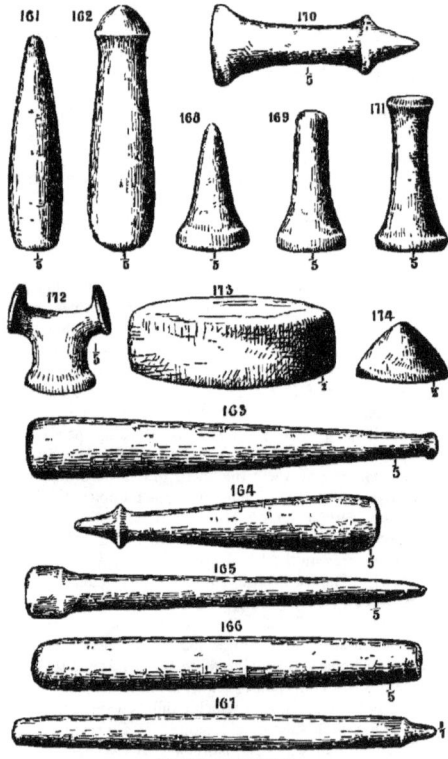

PESTLES AND MULLERS.

five and a half inches in length, and consisting of compact greenstone (Fig. 167) exhibits a somewhat similar character, but is differently shaped at the tapering upper end. It may be of comparatively recent manufacture.

There are short pestles in which the round base is much enlarged, insomuch that the object may be compared to a cone with an inwardly curved side-surface. In many, however, the working part is not convex, but perfectly even, which proves that they were not used in connection with mortars, but were made to operate on a flat surface. Some show, moreover, a small concavity in the centre of the working part, undoubtedly produced by cracking nuts or other hard substances (Fig. 168, greenstone, Pennsylvania; Fig. 169, syenite, Ohio). A very fine specimen from British Columbia, pertaining to the class here described, is encircled by a ring-like projection below the upper end (Fig. 170, greenstone). Another specimen of somewhat recent aspect, and derived from Washington Territory (Fig. 171), is described on the accompanying label as an "Indian hammer used to drive wooden or horn wedges to split wood." This implement consists of a beautiful silicious stone of a light-green color, and is worked with great care.

The most curious and elaborate specimens of the pestle kind were brought from Alaska. These tools are provided with horizontal handles terminating in round plates, slightly convex on the outside. The base or working part in these implements is perfectly even and smooth (Fig. 172, greenstone; apparently an old tool).[21]

Two pestles from Alaska are ornamented at the upper end with rude carvings, representing, respectively, the heads of a bird and of a quadruped, both unrecognizable. A fragmentary pestle from Massachusetts bears at its upper extremity the well-executed figure of a raccoon, and another specimen from the same State is fashioned in imitation of the male organ of generation.[22]

It seems proper to mention in this place more or less carefully worked disc-shaped stones of a size to be conveniently grasped with the hand, which, to judge from the smoothness of their flat faces, were applied in triturating grain or other substances (Fig. 173, greenstone, Georgia). In conclusion, reference should be made to a class of small conoid-shaped mullers, sometimes made of hematite, which may have been used for rubbing paint (Fig. 174, greenstone, Ohio). Specimens of this description are not very abundant.

17. Tubes.—Among the aboriginal relics of somewhat enigmatical character are stone tubes of cylindrical and other shapes and various lengths, which sometimes terminate at one end in a sort of mouth-piece. While the smaller ones, which often measure only a few inches, have been thought to represent articles of ornament, or amulets, different purposes have been ascribed to the larger specimens. Schoolcraft seems to consider these latter

---

[21] The writer was informed by Mr. W. H. Dall, that such pestles were formerly used by the natives of Alaska for mixing berries, fish-oil, fat, etc., in the preparation of an article of food. Such implements are no longer made. A few specimens are still in the possession of the aborigines, who preserve them as heirlooms.

[22] Several carved stone objects of this class, not forming parts of pestles, are in the collection. They seem to indicate a love for the obscene rather than anything like phallic worship.

as telescopic instruments which the ancient inhabitants used for observing the stars. This view, it appears, has been generally rejected. There is more probability that the tubes, in part at least, were implements of the medicine-men who employed them in their pretended cures of diseases. They applied one end of the tube to the suffering part of the patient, and sucked at the other end, in order to draw out, as it were, the morbid matter, which they afterward feigned to eject with many gesticulations and contortions of the body. Coreal, who traveled in America from 1666 to 1697, calls the tubes employed by the medicine-men of the Florida Indians "a kind of shepherd's flute" (*une espèce de chalumeau*).²⁴ They are referred to by Venegas²⁵ and Baegert²⁶ as being in use among the Californians, and the German traveler Kohl saw, as late as 1855, one of the above-mentioned cures performed among the Ojibways of Lake Superior. In this instance, however, the tube used by the medicine-man was a smooth hollow bone, probably of the brant-goose.²⁷

The specimens in the Smithsonian collection chiefly consist of light-gray steatite, of striped slate, or of chlorite. As a typical object (Fig. 175, Tennessee) the writer would mention a beautifully polished cylindrical tube,

TUBES (¼).

measuring nearly six inches in length. The carefully drilled perforation has at one end a diameter of about one-fourth of an inch, but it gradually expands until it reaches at the opposite end a diameter of three-fourths of an inch. The striæ produced by the drilling process are distinctly visible. Another specimen of a different (but not uncommon) type is encircled in the middle by a raised ring, and expands toward the ends (Fig. 176, chlorite, Tennessee). The large cavity is not drilled, but rather irregularly scooped out with a tool from both sides, narrowing considerably toward the middle, where it has a diameter of half an inch.²⁸ The Gosh-Utes of Western Utah use at the present day small pipes of somewhat similar shape, and hence it is not altogether improbable, that the tubes of the type just mentioned were smoking utensils. In fact, that character has been ascribed to various kinds of objects of tubular shape.

---

²⁴ Coreal: Voyages aux Indes Occidentales, Amsterdam, 1722, Vol. I, p. 39.
²⁵ Venegas: History of California, London, 1759, Vol. 1, p. 97.
²⁶ Baegert: Account of the Aboriginal Inhabitants of the Californian Peninsula, in Smithsonian Report for 1864, p. 386.
²⁷ Kohl: Kitschi-Gami, oder Erzählungen vom Obern See, Bremen, 1856, Vol. I, p. 148.
²⁸ A very fine specimen of this class, nearly fourteen inches long, lately has been deposited in the National Museum. It was obtained near Knoxville, Tennessee.

SMITHSONIAN ARCHÆOLOGICAL COLLECTION.   45

A very remarkable tube of striped slate, thirteen inches long, and terminating at one end in a broad mouth-piece, was obtained by Messrs. Squier and Davis in a mound near Chillicothe, during their survey of the aboriginal earthworks in the State of Ohio. This specimen, represented by a cast in the collection (No. 7243), is figured and described on page 224 of the "Ancient Monuments of the Mississippi Valley" by Squier and Davis, forming the first volume of Smithsonian Contributions to Knowledge.

18. Pipes.—No class of aboriginal productions of art exhibits a greater diversity of form than the pipes carved from stone or moulded in clay. Indeed, a volume would be required for figuring and describing the various shapes of these utensils, the manufacture of which offered to the aboriginal artist an unlimited scope for displaying his individual skill and ingenuity. Some of the more marked types only can be noticed in this account. Stone was the material chiefly used in the manufacture of these smoking utensils, though pipes of clay are by no means uncommon.[29] In the following enumeration of typical pipes of earlier date those of clay have been included—somewhat in violation of the plan of arrangement—in order to avoid the necessity of treating them separately in the section relating to the ceramic manufactures of the aborigines.

Numerous stone pipes of a peculiar type were obtained, many years ago, by Messrs. Squier and Davis during their survey of the ancient earthworks in the State of Ohio. They have been minutely described and figured by them in the first volume of Smithsonian Contributions to Knowledge. The originals of these remarkable smoking utensils (presently to be described) are now in the Blackmore Museum at Salisbury, England; but the National Museum possesses casts of them, which enable visitors to become acquainted with their character. These pipes were formerly thought to be chiefly made of a kind of porphyry, a substance, which, by its hardness, would have rendered their production extremely difficult. That view, however, was erroneous; for since their transfer to the Blackmore Museum they have been carefully examined and partly analyzed by Professor A. H. Church, who found them to consist of softer materials, such as compact slate, argillaceous ironstone, ferruginous chlorite, and calcareous minerals.[30] Nevertheless, they constitute the most remarkable class of aboriginal products of art thus far discovered; for some of them are so skillfully executed that a modern artist, notwithstanding his far superior metallic tools, would find no little difficulty in reproducing them.

---

[29] The navigators who first visited the Atlantic Coast of North America noticed copper pipes among the natives, as, for instance, Robert Juet, who served under Hudson as mate in the Half-Moon. Such pipes must be very rare. There are none in the Smithsonian collection.

[30] The subject is fully treated in "Flint Chips," by E. T. Stevens, London, 1870. From this valuable work the drawings of some of the pipes recovered by Messrs. Squier and Davis are here copied, the original woodcuts used in illustrating the "Ancient Monuments of the Mississippi Valley" having been destroyed by the fire which visited the Smithsonian building in 1865. Figs. 117 to 184 are reproductions of illustrations contained in Mr. Stevens' work.

Four miles north of Chillicothe, Ohio, there lies, close to the Scioto River, an embankment of earth, somewhat in the shape of a square with strongly rounded angles, and enclosing an area of thirteen acres, over which twenty-three mounds are scattered without much regularity. This work has been called "Mound City," from the great number of mounds within its precinct. In digging into the mounds, Squier and Davis discovered hearths in many of them, which furnished a great number of relics, and from one of the hearths nearly two hundred stone pipes of singular form were taken, many of which, unfortunately, were cracked by the action of fire, or otherwise damaged. The occurrence of such pipes, however, was not confined to the mound in question, others having been found elsewhere in Ohio, and likewise in mounds of Indiana. In their simple or primitive form they present a round bowl rising from the middle of a flat and somewhat curved base, one side of which communicates by means of a narrow perforation, usually one-sixteenth of an inch in diameter, with the hollow of the bowl, and represents the tube or rather the mouth-piece of the pipe, while the other unperforated end forms the handle by which the smoker held the implement and approached it to his mouth. A remarkably fine specimen of this kind (Fig. 177) was found in a mound of an ancient work in Liberty Township, Ross County, Ohio. In the more elaborate specimens from "Mound City" the bowl is formed in a few instances in imitation of the human head, but generally of the body of some animal, and in the latter cases the peculiarities of the species which have served as models are frequently expressed with surprising fidelity. The human heads, undoubtedly the most valuable specimens of the series, evidently bear features characteristic of the Indian race, and they are further remarkable for the head-dress or method of arranging the hair (Fig. 178). A few of the heads show on the face incised ornamental lines, obviously intended to imitate the painting or tattooing of the countenance. The following mammals have been recognized: the beaver (Fig. 179), the otter (with a fish in its mouth, Fig. 180), the elk, bear, wolf, panther, wild-cat, raccoon, opossum, squirrel, and sea-cow (manati, lamantin, *Trichecus manatus*, Lin.). Of the animal which is supposed to represent the sea-cow, seven carvings have been found. This inhabitant of tropical waters is not met in the higher latitudes of North America, but only on the coast of Florida, which is many hundred miles distant from Ohio. The Florida Indians called this animal the "big beaver," and hunted it on account of its flesh and bones.[31] More frequent are carvings of birds, among which the eagle, hawk, falcon, turkey-buzzard, heron (Fig. 181), several species of owls, the raven, swallow, parrot, duck, and other land and water-birds have been recognized. One of the specimens is supposed to represent the toucan, a tropical bird not inhabiting the United States; but the figure is not of sufficient distinctness to identify the original that was before the artist's mind, and it would not be safe, therefore, to make this specimen

---

[31] Bartram: Travels, Dublin, 1793, p. 229.

the subject of far-reaching speculations. The amphibious animals, likewise, have their representatives in the snake, toad, frog, turtle, and alligator. One specimen shows a snake coiled around the bowl of the pipe. The toads, in particular, are faithful imitations of nature. Leaving aside the more than doubtful toucan, the imitated animals belong, without exception, to the North American fauna, and there is, moreover, the greatest probability that the

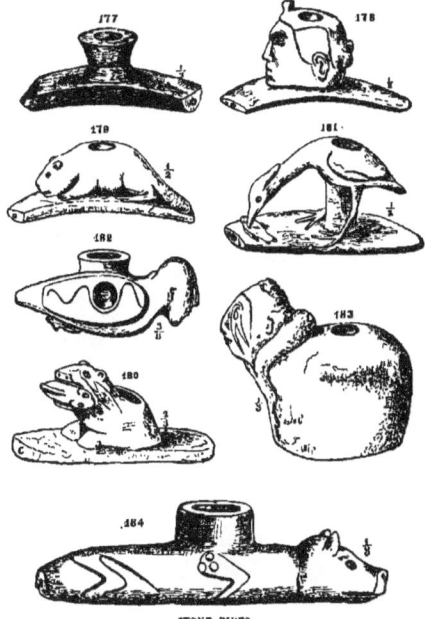

STONE PIPES.

sculptures in question were made in or near the present State of Ohio, where, in corroboration of this view, a few unfinished pipes of the described character have occurred among the complete articles.

Pipes of this type are generally of rather small size, and in many the cavity of the bowl designed for holding the narcotic is remarkable for its insignificant capacity. These pipes were probably smoked without a stem, the narrowness of the perforations in their necks not permitting the insertion of anything thicker than a straw or a very thin reed. Yet most of the pipes of earlier date, occurring either in mounds or on the surface of the ground, are provided with a hole of suitable size for the reception of a stem. A very remarkable stone pipe of this character, obtained during the survey of the Ohio earthworks by Squier and Davis, was found within an ancient enclosure, twelve miles below the city of Chillicothe. It represents the body of a bird with a human head exhibiting strongly marked Indian features (Fig. 182). The original, not having been exposed to the action of fire, is in an excellent state of preservation, and retains its original beautiful polish.

The name "calumet-pipes" has been given to large stone pipes which were smoked with a stem, and are usually fashioned in imitation of a bird, mammal, or amphibian, and sometimes of the human figure. They were thus called on account of their bulk, which seemed to indicate their character as pipes of ceremony, to be used on solemn occasions. It was further thought these pipes had not been the property of individuals, but that of communities, a view which does not seem to be altogether correct, since some have been discovered in burial-mounds, accompanying a single skeleton.

A pipe of the kind just mentioned is made of ferruginous sandstone, and represents rather rudely a human figure with a snake folded around its neck (Fig. 183, cast, Paint Creek, Ross County, Ohio). The face is marked with incised lines. Another large calumet-pipe, carved in imitation of a quadruped of the canine family (probably a wolf), consists of chlorite, and was found in Ross County, Ohio (Fig. 184, cast). The National Museum possesses one of the finest calumet-pipes thus far discovered in the United States. It is boldly cut out of potstone, and represents a bird with a strongly curved beak, perhaps an eagle, which stands on a high pedestal, showing in front an inverted human face bearing incised lines. The bowl rises from the back of the bird. This remarkable aboriginal carving (Fig. 185), which partakes somewhat of a "Promethean" character, and may have reference to an event or to some religious conception, was found in the State of Kentucky.

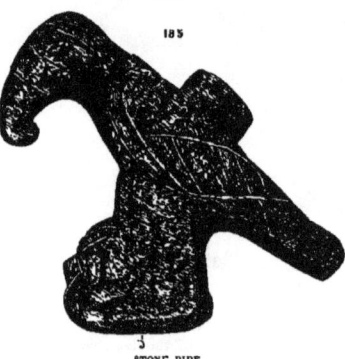

185

STONE PIPE.

There are many small pipes which, though they were smoked with stems, are not provided with projections or necks for their insertion, thus resembling one of the calumet-pipes just noticed (Fig. 183). The holes designed to hold them are drilled immediately into the body of the bowl. Pipes of this description assume innumerable forms. Some are produced without much art, almost reminding one of the corn-cob pipes in use among the farmers of this country (Fig. 186, compact argillaceous stone, Pennsylvania); others are fashioned with great care, and may be counted among the better class of Indian products of art. As an example the writer would mention a highly polished serpentine pipe from West Virginia, which might at first sight easily

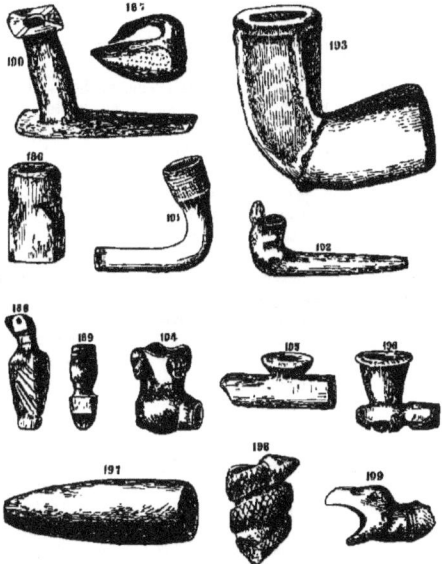

STONE AND CLAY PIPES (½).

be mistaken for the imitation of a swan, though it has been ascertained by competent ornithologists that the bird was intended for the loon (Fig. 187). Generally speaking, birds were rather frequently the models copied by the manufacturers of the pipes under notice, and an upright bird figure, with the receptacle of the narcotic hollowed out downward between the wings and an aperture for the stem at some distance from the end of the tail, may be considered as a typical form. A specimen of this description (Fig. 188), made of a compact argillaceous stone and representing a parrot, was derived from the State of New York. In another class of pipes, somewhat analogous to the preceding type, the barrel-shaped bowl rises from a sort of handle pierced with a hole for the stem (Fig. 189, argillaceous stone, Ohio). Pipes of this character may not be very old.[27]

Passing over to the pipes provided with necks, a typical class deserves mention, in which the almost cylindrical very high bowl stands upon a flat perforated base prolonged beyond the bowl, to form a sort of handle. In some the perforation of the neck is very narrow, and these were probably smoked without stems, like the pipes obtained from mounds in Ohio, to which they bear some analogy (Fig. 190, Virginia). The specimens of this description seen by the writer were made of chlorite.

In the districts formerly inhabited by the Iroquois tribes, and in the neighboring parts, there have been found pipes of stone and clay in which the connection of the bowl with the neck forms a curve (Fig. 191, serpentine, New York). Some of these pipes, more especially specimens of burned clay, exhibit elegant outlines, almost reminding one of a cornucopia. The length of the neck in some of the specimens and their narrow bore seem to indicate that they were smoked without separate stems, like the common clay pipes now in use, in which bowl and stem are united. A very beautiful, highly polished steatite pipe of the collection is carved in imitation of a lizard (Fig. 192, Pennsylvania). The straight neck or stem apparently forms the animal's tail, and its toes are indicated by incised lines.

Many of the pipes formerly used by the aborigines, and made either of stone or clay, approach in general character certain pipes common among civilized races, being furnished with distinct necks by which they were attached to stems. Such pipes are often of large dimensions, and their bowls provided with wide cavities for holding a considerable quantity of the smoking material. These large specimens, or calumets, nearly always consist of stone, and their bowls and necks are round or four-sided in the cross section (Fig. 193, potstone, North Carolina). Incised lines, raised rims and other ornaments, characterize the more elaborate specimens of this kind. A beautiful serpentine pipe of smaller size, and, perhaps, not very old, shows a quadrilateral rim with a human head carved at each corner (Fig. 194, Texas). Several small stone pipes of the collection are remarkable for their low broad-rimmed bowls and the prolongation of the necks beyond the bowls. One of the specimens of this character, which consists of compact limestone, is evidently very old, being entirely covered with a white crust produced by decay (Fig. 195, mound in Kentucky).[28]

Clay pipes of kindred character, moulded into almost every conceivable shape, frequently occur in aboriginal graves as well as on the surface. The

---

[27] The type occurs among the pipes carved by modern Indians.
[28] The pipes made of red pipestone or Catlinite, which are represented by numerous specimens in the collection, belong to more recent times.

bowl often represents a more or less carefully executed human head. Some bear some resemblance to the *chibouc* of the Turks (Fig. 196, Georgia).

Stone pipes of an altogether different character were in vogue among certain Californian tribes. They are of an elongated conoidal shape, of large size and corresponding capacity (Fig. 197, serpentine, Santa Barbara County). Some have been found with a short hollow bone cemented as a mouth-piece into the aperture at the tapering end.[34] Similar pipes of smaller size are still used by the Pai-Utes.

Lastly, special mention should be made of two fragmentary pipes of clay, both found in Madison County, New York, and remarkable for excellent workmanship. In one the bowl is formed by the coils of a skillfully executed snake (Fig. 198); in the other by the head of a bird (apparently a raven) with widely opened bill (Fig. 199). The outside of these specimens is coated with a yellowish brown paint, and perfectly smooth.

19. Ornaments.—Though the aborigines of North America (north of Mexico) chiefly employed shell-matter as the material of their ornaments, they likewise made use of stone for that purpose. First ought to be mentioned their stone beads of various forms and sizes, which they strung and wore as personal decorations, mostly, perhaps, in the shape of necklaces. Some beads are globular or compressed at the opposite ends (Fig. 200, serpentine, Santa Barbara County, California); others are of irregular shape, four-sided, notched at the circumference, etc. (Figs. 201 and 202, potstone, Pennsylvania). The collection contains a number of articles of ornamental character, presenting the shape of straight tubes, either cylindrical or somewhat swelling toward the middle. A well-drilled specimen consisting of silicious material (Fig. 203, Mississippi) measures nearly three inches in length. There are further in the collection several ornaments made of the red pipestone, or Catlinite, from the Coteau des Prairies in Minnesota. Though probably no great antiquity can be ascribed to them, they ought to be mentioned here, having been discovered in digging the Oriskany Canal in the State of New York. They may be attributable to the Iroquois. A typical form of these ornaments, which the writer had occasion also to notice outside of the National Museum, may be likened to a compressed slender pyramid, pierced in the longitudinal direction (Fig. 204). The occurrence of these objects of Catlinite in the State of New York, distant twelve or thirteen hundred miles from the red pipestone quarry, furnishes a strong evidence of a far-extended aboriginal trade.

Next must be mentioned objects of stone pierced for suspension, which were undoubtedly worn as breast ornaments, representing in many cases, it may be assumed, badges of distinction. A very fine specimen of the collection, somewhat resembling in outline a certain class of pierced tablets,

---

[34] Among the objects recovered by Mr. Paul Schumacher, during his explorations in Southern California, are many pipes of this description.

is ornamented with a border of dotted triangles (Fig. 205, trap rock, Connecticut). Another smaller specimen of kindred character, which is made of a flat sandstone pebble of oval outline, bears incised ornamental lines (Fig. 206, Rhode Island). Small oval or round pebbles of a compressed form, pierced with a hole for suspension, but otherwise left in their natural

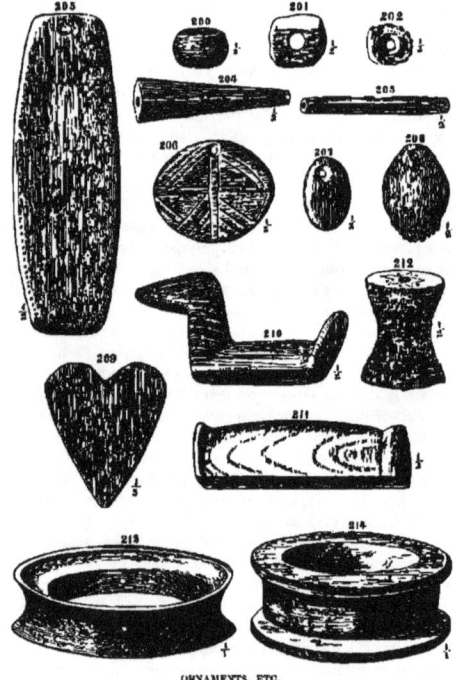

ORNAMENTS, ETC.

state, are not uncommon (Fig. 207, Pennsylvania). A very remarkable small object, designed for suspension, consists of a shell of brown hematite of rather irregular outline, and shows no other modifications by art than a perforation at one extremity, and nine distinct incisions or notches at the other (Fig. 208, Virginia). These notches may have a significance, denoting, perhaps, the number of animals of a certain species, or of enemies, killed by the wearer. Possibly the notches may form the enumeration of transactions of a more peaceable character. Similar "records" have been noticed among the prehistoric relics of Europe. Several objects in the collection, undoubtedly ornamental in their character, are shaped like a heart, which was among the North American Indians, as well as with other nations, the emblem of courage and of other manly qualities. One of the specimens (Fig. 209, argillaceous slate) is derived from an Ohio mound, where it was lying near the neck of the skeleton.

The Smithsonian collection possesses a series of the well-known curious relics intended to represent birds, the body, neck, head, and tail being clearly, though clumsily, indicated. The place of the eyes is occasionally marked by small bead-like elevations, sometimes, however, by unproportionally large knob-shaped projections. These figures stand on flat bases pierced at their extremities with diagonal holes which often exhibit traces of wear. The objects are generally made of soft stone, such as the often-mentioned green striped slate; yet porphyritic syenite and other hard substances sometimes form their materials. A characteristic specimen of the collection (Fig. 210, Pennsylvania) consists of striped slate. In this instance the eyes are indicated by small round protuberances. The object is marked along the neck, head, back, and tail with numerous notches, probably designed for ornament. It is evident that these relics were worn in some way (perhaps as amulets), and not used as knife-handles or for removing the husk of Indian corn, as has been suggested. The latest theory, based upon information received from an "aged Indian" is, that they were worn in olden times on the heads of Indian women, but only after marriage. One specimen of the collection, however, made of striped slate, and finished in every respect, though left without the diagonal holes, weighs two pounds and one ounce. It is incredible that a woman should have worn such a heavy object on her head for the sake of indicating her married state. Some very fine specimens in the National Museum, evidently belonging to the class under notice, are not imitations of the bird form, but are shaped alike at both perforated extremities (Fig. 211, striped slate, mound in Ohio).

The objects hitherto treated may be denominated ornaments with some degree of safety; but we include here, for want of a better designation, a few other classes of typical articles which possibly were intended for purposes of a useful rather than a decorative character. Among them are small carefully worked objects shaped like cylinders with inwardly curved side-surfaces. These objects are perforated lengthwise, or show at least the beginnings of perforations at both ends, and bear on their side-surfaces incised ornamental lines. The round (sometimes oval) flat extremities are likewise ornamented with engraved lines and dots, differing in their pattern from the decorations on the curved sides (Fig. 212, fine-grained argillaceous sandstone, Kentucky). The mode of their application has not been ascertained. It has been sug-

54   PECKED, GROUND AND POLISHED STONE.

gested, on account of their concave side-surfaces, that they were tools employed in pressing ornamental lines on clay vessels while yet in a soft state. Upon trial, however, it has been found that the impressions produced by them on wet clay bear little analogy to the ornamentation which characterizes North American vessels, and hence their real purpose remains problematical for the present.

Among the relics of the former population are rings of stone and bone of different sizes, but similar in shape, being deeply grooved upon the outer edge, and pierced with eight equidistant small holes radiating from the centre. A cast in the collection (Fig. 213) is the fac-simile of such a ring, which was discovered in a mound not far from Chillicothe, Ohio. The cast, however, represents the object as perfect, whereas the original, formerly belonging to Dr. E. H. Davis, constitutes only one-half of the ring, which consists of a dark stone of medium hardness. In a former publication[26] the writer has suggested that these rings once formed parts of bow-drills by means of which the aborigines produced the perforations in pipes and other objects of stone. A well-made potstone ring of the collection (Fig. 214, Pennsylvania) is grooved around the circumference, but not pierced with lateral holes. The writer's view concerning the mode of application of these rings has been somewhat shaken by the fact that there is in the Smithsonian collection a similarly shaped ornamented ring of burned clay, which, owing to the fragility of its material, hardly could have been utilized in the indicated manner. Yet this clay ring, though resembling the described objects of stone, may have been designed for a totally different purpose.

20. **Sculptures.**—Though many of the objects treated in the preceding portion of this account may be called sculptures in view of the mode of their production, as for instance, stone pipes and other elaborately wrought articles of the same material, the expression is here reserved for a special class of aboriginal relics, among which imitations of the human body, or parts of it, are the most conspicuous.

There are in the collection numerous casts of Mexican stone masks and images, some of which probably have reference to the idol-worship of the Aztecs. The so-called masks are not uncommon in the United States, and casts of several of them may be seen in the collection, which also contains an original of this class from Rhode Island, representing a human face very rudely carved in sandstone. The eyes are represented by oval depressions, and a simple groove constitutes the mouth, while the nose is indicated by an insignificant elevation. The back part shows a rough fracture, a circumstance which renders it probable at least that the specimen is the detached facial portion of a very roughly worked imitation of the human head.

One of the most valuable objects in the National Museum (Fig. 215) is a

---
[26] "Drilling in Stone without Metal;" Smithsonian Report for 1868.

stone image, more than twenty inches in length and weighing thirty-seven pounds and four ounces, which was discovered in a cave near Strawberry Plains, sixteen miles east of Knoxville, Tennessee.[26] This remarkable relic, which is in a very good state of preservation, consists of crystalline limestone, the fracture of which can be seen at the back of the head, where the figure seems to have been detached from the rock out of which it was sculptured.

SCULPTURES.

It is possible, however, that the fracture indicates the former presence of some sort of handle or projection by which the image was carried or attached. The face shows a somewhat prominent nose and strongly marked brows, and the eyes consist of small oval cavities, while the mouth is ring-shaped, as in many Mexican representations of the human countenance. A groove extends across

---

[26] An account of the discovery of this image is given in the Smithsonian Report for 1870, p. 385.

the face between the nose and the mouth. The ears are unproportionally large. There is no body, properly speaking, but merely a kind of four-sided pedestal with a flat base on which the figure can stand. Its front side shows an appendage in the form of a small apron, which may, however, be intended to mark the male sex. Lastly, there are to be seen on both sides of the figure cavities, perhaps cut out in lieu of arms. The stone image just described is undoubtedly among the best of its kind thus far discovered within the United States, and compares favorably with kindred sculptures of Mexican or Central American origin.

The sculpture of a human head, cut almost in life size from a kind of limestone, is of interest, irrespectively of its intrinsic value, on account of having been in the possession of President Thomas Jefferson, while he lived at Monticello (Fig. 216). Although much mutilated, this relic is still sufficiently preserved to show the very creditable original workmanship. There is no exaggeration or deformity in any part of this head, which may be the likeness of some aged person with a deeply wrinkled face. A conical cavity in the base of the head evidently served for keeping it in position by some sort of support. There is another cavity in the back part of the head. The records of the Smithsonian Institution contain no information as to the place where it was found.

A curious little relic, made of a dark ferruginous stone, deserves notice on account of its grotesque character (Fig. 217, Ohio). The stone seems to be a natural formation, only modified by the carving of round eyes, a nose, and a wide open mouth.

It is well known that the Mexicans were far more advanced in the art of stone sculpture than the Indian tribes inhabiting higher latitudes of North America. There are in the collection some remarkable specimens of Mexican stone sculpture, among them a massive slab worked in the shape of a human head surmounted by an elaborate head-dress (Fig. 218). This relic, obtained from Tuspan, consists of some kind of volcanic rock, and may have belonged to a large figure. The head measures fifteen inches in length and is thirteen inches and a half broad. Small Mexican carvings in stone are not wanting in the National Museum; but as a description of all these specimens would occupy too much space, only a few will be noticed. There is, for instance, a flat carving of the human figure, in which the head alone, including a peculiar head-dress, is carefully, though not artistically, executed in its details, while the body merely forms a sort of appendage (Fig. 219). This relic consists of a greenish-gray stone, but not of the much-valued *chalchihuitl*. Another small specimen, measuring about an inch and a half in height, and carved from white alabaster, represents a human figure with a remarkable countenance and an unproportionally small body in the squatting posture characteristic of Mexican images (Fig. 220). The neck is pierced for suspension. Lastly, we would mention a carving in the shape of a death's head, not larger than a walnut, which was found among the ruins of Chichen Itza, in Yucatan.

The flat back of this diminutive representation of a skull is perforated at each side with a diagonal hole. The material appears to be silicified wood.[*]

A very curious class of Indian sculptures are the imitations of human footprints which occur, cut on solid rocks and sometimes on boulders, in various parts of North America. These artificial tracks have elicited much unprofitable speculation, being considered by some as real impressions of human feet, and consequently dating from a time when the rocks were still in a state of softness. Though this view is now entirely abandoned by intelligent observers, there are some persons who, being unacquainted with the results of geology, still adhere to it. The foot-prints of man which are found, either isolated or in connection with other designs, on many rocks in the United States belong to the pictographical system of the aborigines, and probably relate to incidents worthy of their remembrance. Among the remarkable objects of the collection are three large stone slabs bearing impressions of human

SCULPTURED FOOT-TRACKS (⅛).

feet. On two of these slabs, which have been carefully cut out of the rocks, may be seen, respectively, two impressions of feet represented as being covered with moccasins of a pattern still in use among the Sioux and other western tribes (Fig. 222). These slabs consist of sandstone, probably pertaining to the carboniferous formation, and were obtained from the banks of the Missouri River. The third specimen of this description (Fig. 223) is a flattish block of quartzite (probably a boulder), which bears on one of its flat sides the impression of a naked foot, each toe being distinctly marked by a cavity of proportionate depth. The foot is surrounded by a number of cup-shaped depressions. This relic was obtained in Gasconade County, Missouri.

---

[*] This relic is described in the Smithsonian Report for 1871, p. 423.

Though perhaps not exactly in the right place, we would here mention a specimen of the collection, which was evidently employed in the manufacture of moccasins, being, in fact, a stone brought to the shape of a last, such as shoemakers use. This specimen, consisting of greenstone, was obtained in Arizona; but similar stone lasts also occur in the eastern parts of the United States. There is one in the collection of the Antiquarian Society at Worcester, Massachusetts, and the writer has seen specimens of the same character in Missouri and New Jersey.

II. COPPER.

It is well known that the North American Indians, at least those inhabiting the districts north of Mexico, lived in an age of stone at the time when their country became first known to the whites. They made, however, some use of native copper which they chiefly obtained from the region where Lake Superior borders on the northern part of Michigan. The traces of ancient aboriginal mining in that district were first noticed in 1847, and since that time the subject has been fully treated in various publications, more especially in a memoir by Mr. Charles Whittlesey, forming one of the Smithsonian Contributions to Knowledge.[1] Native copper from other parts of the United States likewise may have been utilized to some extent by the aborigines.[2] Copper implements, such as axes or celts, chisels, gravers, knives and points of arrows and spears, together with ornaments of various kinds, have been found in mounds and on the surface in different parts of this country, though not in great abundance, and it does not seem, therefore, that copper played an important part in the industrial advancement of the race. The aborigines lacked, as far as investigations hitherto have shown, the knowledge of rendering copper serviceable to their purposes by the process of melting, contenting themselves by hammering masses of the native metal with great labor into the shapes of implements or of objects of decoration. In short, they treated copper as malleable stone. Copper articles of aboriginal origin generally exhibit a distinct laminar structure, though quite a considerable degree of density has been imparted to the metal by continued hammering. It must be admitted, furthermore, that the natives had acquired great skill in working the copper in a cold state.[3] The first voyagers who visited North America (Verazzano, the Knight of Elvas, Captain John Smith, Robert Juet, and others) saw copper ornaments and other objects made of this metal in the possession of the Indians, and there can be little doubt that the manufacture

---

[1] "Ancient Mining on the Shores of Lake Superior," Washington, 1863.

[2] It is sometimes met, in pieces of several pounds' weight, in the valley of the Connecticut River, and also in the State of New Jersey, probably originating in both cases from the red sandstone formation. Near New Haven, Connecticut, a mass was found weighing ninety pounds.

[3] Mr. J. W. Foster describes and figures in his "Prehistoric Races of the United States," North American copper implements, which, as he thinks, were produced by casting. The subject will require further investigation.

of such articles was still going on at the time of the discovery of the North American continent.[4]

The objects of copper found in the United States, as mentioned, embrace implements and weapons as well as ornaments, all of which are represented in the collection by originals and a number of copper casts. First should be noticed the celt-shaped objects, which bear a great resemblance to corresponding bronze implements in European collections. There is, for instance, a well-shaped celt derived from a mound near Lexington, Kentucky, which has been exposed to the action of fire, as seen by pieces of charcoal and cinders still adhering to it (Fig. 224). The implement and the cinders are covered with green rust. From the same mound were taken some axe-shaped, though perfectly blunt objects, terminating at the broader end in lateral curved appendages (Fig. 225). Their significance has not yet been ascertained. Among the copper celts of the collection are several smaller specimens of good workmanship, one of them (Fig. 226) taken from a mound near Savannah, Tennessee. The most beautiful article of a wedge-like character is a kind of chisel with an expanding, strongly curved edge, which shows a slight concavity, imparting to the implement almost the character of a gouge (Fig. 227, back view, New York). The upper surface is nearly even, but the back part presents, as it were, two faces, which join in the middle, forming a longitudinal ridge.

There are further to be mentioned weapons of the arrow and spear-head form, of elongated shape, and terminating opposite the points in stems, either truncated or pointed (Fig. 228, Lake Superior district; Fig. 229, Vermont). A well-made crescent-shaped implement with a tolerably sharp convex edge may be considered as a knife (Fig. 230, Fond du Lac, Wisconsin). If it had been a gorget, as has been suggested, it probably would show the usual holes for suspension. One of the most interesting copper tools of the collection, perhaps a unique relic, is a slender awl still inserted in its bone handle (Fig. 231). This specimen, which was found on Rhea's Island, Loudon County, Tennessee, reminds one of corresponding iron tools in use at the present day. A copper sinker from Ohio (Fig. 232), analogous in shape to a certain class of stone objects previously described, deserves particular notice.

Passing over to the copper ornaments of the collection, we will first mention

---

[4] Traces of wrought silver have been discovered among the aboriginal relics, but they are so exceedingly scanty that the technical significance of this metal hardly can be taken into consideration. Native silver, it is well known, occurs interspersed in small masses in the copper of the Lake Superior district, and from that source the Indians doubtless derived the small amount of silver used by them. Gold was seen by the earliest travelers in small quantities (in grains) among the Florida Indians; yet, to the writer's knowledge, no object made of gold, that can with certainty be attributed to the aborigines (north of Mexico), has thus far been discovered. Squier and Davis found no gold during their extensive explorations in Ohio. The discovery of small aboriginal relics of gold, however, would not be surprising, considering that this precious metal occurs in some of the districts of the United States formerly occupied by the Indian race.

It is not probable that the natives understood the melting of lead; but pieces of galena frequently occur in mounds, and there is in the Museum a bead (resembling the original of fig. 259) skillfully made of that ore.

armlets and bracelets, consisting of hammered rods about the thickness of a lead-pencil, and bent into a circular or oval form until their ends meet. These specimens were obtained from mounds in Indiana and West Virginia. Similar

COPPER IMPLEMENTS AND ORNAMENTS (⅓).

bronze ornaments are met in collections of European antiquities. Copper beads are well represented in the collection. They consist of coarse wire or small pieces of copper closely wound and hammered together (Fig. 233, mound in Ohio), or, more generally, of strips of copper sheet bent into the form of cylinders with overlapping, though never soldered, edges. These cylindrical beads are sometimes so long that they may be called tubes, as, for instance, a number of specimens more than three inches long, which were discovered in an Indian grave near Newport, Rhode Island (Fig. 234). These tubular ornaments, however, though covered with verdigris, cannot be very old, considering that each of them encloses a tightly fitting piece of reed of equal length, evidently stuck into the cylinders for diminishing the width of the holes, and even remnants of the narrow thong by which they were connected or attached have been preserved. It is probable that the tubes are of Indian (not European) workmanship, and their appearance bears witness to a comparatively recent origin. For aught we know, the wearer may have been a contemporary of Roger Williams.

Among the copper finds in various parts of the United States have been noticed curious small objects somewhat resembling spools in shape, consisting of two concavo-convex discs connected by a central hollow axis. Objects of this class are said to have been discovered with thread wound around the axis. The collection contains a number of such relics, most of which were derived from mounds near Savannah, Tennessee (Fig. 235). Their use has not yet been explained. From a mound in the same neighborhood was obtained a piece of copper sheet resting on a fragment of much decayed bark or grass matting, impregnated with the green rust of the copper (No. 9882 of the collection).

Farther to the north, in Alaska, some of the aboriginal tribes have long been known to employ in the manufacture of tools and weapons native copper obtained from a locality on the Atna or Copper River, where it occurs in rolled masses, sometimes weighing thirty-six pounds. Copper articles made by natives of Alaska may be seen in the ethnological department of the Museum.

III. BONE AND HORN.

Although, generally speaking, implements of bone and horn of early date are not very abundant in North American collections, they are represented by many characteristic specimens in the National Museum, the objects of bone outnumbering those of horn. The teeth and claws of wild animals, it will be seen, were chiefly made into ornaments testifying the valor of their wearers. Piercers obtained from mounds, shell-heaps, etc., form the most numerous class of bone tools (Fig. 236, ancient village site on one of the Aleutian Islands; Figs. 237 and 238, mounds in Union County, Kentucky). These perforators bear a striking resemblance to those found among the relics of the ancient lake-men of Switzerland. A beautiful bone needle of somewhat recent appearance deserves special notice (Fig. 239, San Miguel Island, California). This needle is not pierced with an eye, but exhibits in its stead two grooves for fastening the thread. There are in the collection several bone harpoon-heads, barbed on one side, and pierced with a hole for attachment (Fig. 240, grave in Michigan; Fig. 241, Alaska).[1] Somewhat similar armatures of bone, derived from the caves of the Dordogne, in Southern France, are described by Lartet and Christy in the "Reliquiæ Aquitanicæ." Speaking of fishing implements, we would mention well-wrought bone hooks from Santa Cruz Island, California (Fig. 242). The shanks of the hooks are still covered with a coating of asphaltum, evidently applied for securing the line. Contrary to the general rule, the barbs in these hooks are placed on the outer side.

California, further, has furnished a number of whistles apparently made of bird bones and provided with a blowing-hole not in the middle, but placed nearer one extremity of the hollow bone than the other (Figs. 243 and 244, Mare Island). Other curious objects derived from California, more especially from the Santa Barbara Islands, are cups very ingeniously hollowed out from the vertebræ of cetaceans (Fig. 245, Santa Cruz Island). These cups are partly filled with asphaltum, apparently prepared to serve as paint.[2]

---

[1] Copper harpoon-heads of the same shape (barbs on one side, hole for attachment) may be seen in the collection. They were obtained from Alaska, and belong to the modern fishing gear of the natives.

[2] Since the above was written, the collection has been enriched with many articles of bone and horn, obtained from Californian graves by Mr. Paul Schumacher. Among them we mention large wedges of elk horn and whalebone, polishing tools resembling paper-folders, rather ponderous knife-shaped articles of whalebone, and, lastly, fifes with four holes.

Numerous relics of bone and horn, collected by Mr. F. H. Cushing in the State of New York, lately have been added to the collection. They comprise perforators of various forms and sizes, harpoon-heads, detached prongs of deer horn, more or less polished at their points, and probably employed as smoothing tools, modified beavers' teeth, and various other objects.

64   BONE IMPLEMENTS AND ORNAMENTS.

Like other races of hunters, the aborigines of North America were in the habit of perforating the teeth of wild animals they had killed, and of wearing them as trophies in the shape of necklaces or pendants. The teeth of bears, it seems, formed the most favorite ornaments of this kind, being either left in their natural state and merely pierced at the root (Fig. 246, New York), or

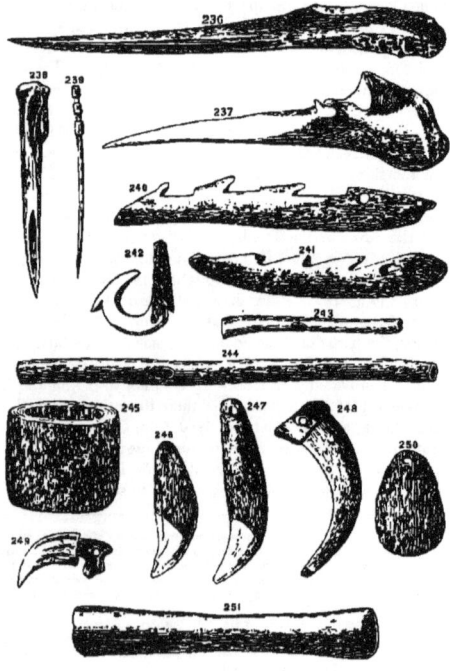

IMPLEMENTS AND ORNAMENTS OF BONE (⅓).

brought into a more regular shape by grinding and smoothing, like a number of specimens from Alaska (Fig. 247), which may, however, belong to a comparatively recent period. Modern Indians, it is well known, wear as tokens of their prowess necklaces made of the claws of the grizzly bear (Fig. 248, Rocky Mountains, recent), and a number of drilled claws of the panther

(*Felis concolor*), derived from California, were probably used in a like manner by the aborigines of that part of North America. In these specimens the perforation passes through the bony part (last phalanx) of the animal's claw (Fig. 249, Santa Cruz Island). A very curious ornament (?) is the pierced epiphysis of a long bone of some animal, probably a deer (Fig. 250, mound in Kentucky).

Besides the objects or classes of objects of bone thus far mentioned there are other specimens, either in a fragmentary state or entire, the purposes of which have not yet been explained. As an example may be selected a wrought hollow bone somewhat expanding at both ends (Fig. 251, Santa Cruz Island, California). It is not attempted to assign a name to this object, which may have been a receptacle or a part of a tool, an unfinished whistle, or, perhaps, an appendage to the dress. There is a possibility, too, that it was the sucking-instrument of a medicine man, made to replace one of the stone tubes which are known to have been employed among the Californians in curing the sick.

IV. SHELLS.

Shells being above other natural productions particularly fitted to be made into ornaments, it is not surprising that they were employed for that purpose by primitive man of all ages and in all parts of the world. The North American tribes utilized, to a great extent, the shells of the sea-coast as well as those of their rivers, and there can be no doubt that marine shells formed an article of exchange in former times, considering that they have been met among aboriginal relics far in the interior of the country. This kind of traffic has been taken up at a later period by white traders, who derived great profit in selling fine specimens to the tribes inhabiting the inland. It is known that the Indians sometimes paid for a fine shell far to the value of thirty or forty dollars, and more. Shells even seem to have been looked upon with a kind of superstitious reverence, and indications are not wanting that they sometimes played a part in their religious ceremonies. Shells, however, were not exclusively converted into ornaments, or preserved as objects of value, but were also employed as utensils, more especially as vessels, an application for which large species, such as *Cassis* and *Bysicon* seemed particularly adapted. The Florida Indians, when first seen by Europeans, used large shells as drinking-cups, and when a chieftain died, the shell which he had used during lifetime (*crater e quo bibere solebat*) was placed on the apex of the mound that marked his place of burial.[1] A large *Bysicon perversum* (*Pyrula perversa*) made into a drinking vessel by the removal of the inner whorls and other modifications may be seen in the collection (Fig. 252, mound in Indiana). Valves of *Unio*-shells, somewhat altered by art, in order to be handled with greater convenience, formed very serviceable spoons (Fig. 253, mound in Kentucky). There are several utensils of this kind in the National Museum. Among other objects designed for useful purposes should be mentioned celts or adzes made of heavy shells, and identical in shape with corresponding tools of stone. Such shell implements have been found on the southern coasts of the United States, especially in Florida, but also at a considerable distance from the sea-board (Fig. 254, Florida; Fig. 255, Kentucky). It further appears that the Florida Indians applied shells of the *Bysicon perversum* as clubs or *casse-têtes* by adapting them to be used with a handle, which was made to pass transversely through the shell. This was effected by a hole

---

[1] De Bry, *Brevis Narratio* (Vol. II, Frankfort on the Main, 1591), Plate 40.

pierced in the outer wall of the last whorl in such a manner as to be somewhat to the left of the columella, while a notch in the outer lip, corresponding to this hole, confined the handle or stick between the outer edge of the lip and the inner edge of the columella. The anterior end of the canal, broken off until the more solid part was reached, was then brought to a cutting edge, nearly in the plane of the aperture. A hole was also made in the posterior surface of the spire behind the carina in the last whorl, evidently for receiving a ligature by means of which the shell was more firmly lashed to the handle.

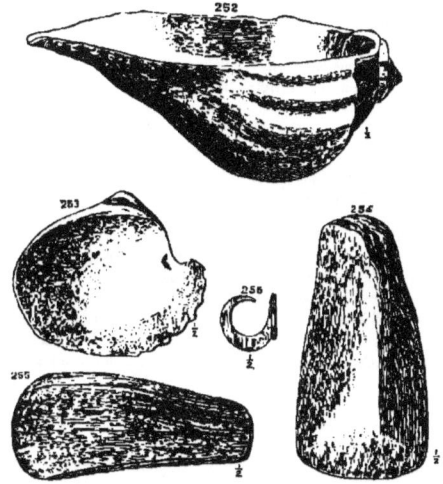

UTENSILS OF SHELL.

Shells prepared in this manner have been found on the shore of Sarasota Bay in Florida, a locality where stone for manufacturing axes is wanting. There are several of these modified shells in the collection.

California has furnished a number of well-wrought fish-hooks, made of the shell of *Haliotis*, which strongly resemble articles of the same description in use among the islanders of the Pacific (Fig. 256, Santa Cruz Island). The Californian coast-tribes, it should be stated, employed shells in various ways, chiefly, however, in the production of objects of personal adornment, which will be considered hereafter. That they utilized the unaltered shells of *Haliotis*, *Cardium*, *Pecten*, *Patella*, *Spondylus*, and *Panopœa* as the receptacles for asphaltum (paint?) is demonstrated by a number of shells still filled with that substance, which were obtained from graves on the Santa Barbara group of islands, and but lately added to the collection of the National Museum.

The Indian shell ornament in its simplest form consists of entire marine shells, such as species of *Marginella, Natica, Pecten, Oliva, Strombus*, etc., and of valves of fresh-water mollusks (chiefly of the *Unio* kind), which, after being truncated at the apex, or pierced with a hole, could be strung together (forming necklaces, bracelets, etc.) or suspended at once without further preparation (Fig. 257, *Strombus pugilis*, shell-heap, Florida;[2] Fig. 258, *Unio*, Tennessee; Fig. 259, *Olivella biplicata*, San Miguel Island, California; Fig. 260, *Oliva literata*, Florida; Fig. 261, *Pecten concentricus*, Florida).[3] Far more frequent than entire shells pierced to be used as beads and pendants are objects of the same class cut from the valves of marine and fluviatile mollusks. The wrought beads exhibit various forms and sizes, but are very frequently found in the shape of more or less regular sections of cylinders, more rarely of prisms, pierced through the middle (Figs. 262 and 263, originals partly covered with oxide of iron, Dos Pueblos, California; Fig. 264, Santa Barbara County, California; Fig. 265, Dos Pueblos). Many shell beads, however, are not cylindrical, but of roundish or irregular contours. The largest beads were made from the columellæ of massive shells (*Bysicon, Strombus*) and many of these still exhibit a portion of the columellar spiral groove (Fig. 266, Georgia). Such beads are generally more or less cylindrical, or globular, and drilled in the direction of the longer axis. Some taper at both ends, resembling a cigar in shape. Very remarkable specimens of this kind were obtained from California (Fig. 267, San Miguel Island). In some of them the spiral groove is deepened by art and filled with asphaltum, doubtless with a view to improve their appearance. Specimens of this kind may have constituted some part of the head-dress.

The aborigines also made from the columellæ of large marine univalves peculiar pin-shaped articles, consisting of a more or less massive stem which terminates in a round knob or disc (Fig. 268, Florida). There have been found specimens measuring six inches in length. Their destination is as yet unexplained; they were, perhaps, attached to the head-dress, or worn as ornaments in some other way.

Of particular interest are the wampum-beads, which represented the *money* among many tribes of aborigines, forming also valued ornaments (necklaces, bracelets, etc.), and constituting the strings and belts of wampum,[4] which played such a conspicuous part in Indian history, being exchanged at the

---

[2] This shell is pierced with a second hole noticeable in the drawing. The size of the holes and the weight of the shell render it possible, that it was not used as an ornament, but for some other purpose. It may have been a net-sinker.

[3] Similarly pierced shells of *Pecten* are strung together and used as rattles by the natives of the Northwest Coast.

[4] The word "wampum" is derived from *wompam*, signifying *white* in the language of the Narragansetts.—Roger Williams: "A Key into the Language of America," Providence, 1827, p. 130.

conclusion of peace and on other solemn occasions, in order to ratify the transaction, and to perpetuate the remembrance of the event. The term "wampum" is often applied to shell beads in general, but might with propriety

SHELL ORNAMENTS (½).

be confined to a certain class of cylindrical beads, usually one-fourth of an inch long, but sometimes much longer, and drilled lengthwise, which were chiefly manufactured from the shells of the common hard-shell clam (*Venus mercenaria*). This bivalve, occurring, as every one knows, in great abundance on the North American coasts, formed an important article of food of the Indians living near the sea, a fact demonstrated by the enormous quantities of castaway clam-shells, which form a considerable part of North American shell-heaps. The natives used to string the mollusks and to dry them for consumption during winter. The blue or violet portion of the clam-shells furnished the material for the dark wampum, which was held in much higher estimation than that made of the white parts of the shells, or of the spines of certain univalves. Roger Williams states that the Indians of New England manufactured white and dark wampum-beads, and that six of the former and three of the latter were equivalent to an English penny.[5] Yet, it appears that even at his time the colonists imitated the wampum, a practice which assumed the proportions of a regular business in later times, porcelain, glass, and enamel being the materials employed in facsimilizing them. Much wampum, however, was made by whites from clam-shells, and hence arises the difficulty of singling out the genuine Indian manufactures. In the intercourse of the New England colonists among themselves, wampum served at certain periods instead of the common currency, and the courts issued, from time to time, regulations for fixing the value of this shell-money. In transactions of some importance it was measured by the fathom, the dark or blue kind generally being double the value of the white.

There are many beads and strings of wampum in the collection; yet owing to the circumstances just mentioned, it is no easy matter to recognize the real Indian productions (Fig. 269, Upper Missouri; doubtless brought there by way of trade). The peculiar kind of wampum here treated was chiefly in use east of the Mississippi River, though shells, either entire or cut into beads, assumed the character of money in parts far beyond that river. Among the tribes of the northwestern coast of North America, from the northern border of California far upward into Alaska, the shells of the *Dentalium* represented, until within the latest time, the wampum of the eastern regions, being used, like the latter, both as ornament and money. These shells, which occur in certain places of the Pacific coast, may be likened to small, tapering, and somewhat curved tubes, and, being open at both ends, they could be strung without further preparation. Among the Southern Californians the circulating medium consisted, according to H. H. Bancroft, of small round pieces of the white muscle-shell. These were perforated and arranged on strings, the value of which depended on their length. There is a quantity of small perforated shell discs in the collection, which were obtained from Southern California, and may have constituted the money of the aborigines. These small discs, however, are concavo-convex, and evidently were not cut from the muscle-shell.[6]

---
[5] A Key, etc., p. 128.
[6] It appears probable that among the natives of that region the *Olivella biplicata* and the land-shell *Helix strigosa* served as substitutes for money. Mr. Paul Schumacher discovered on San Nicolas Island deposits of these shells, which had been stored in the sandy ground, and formed diminutive hillocks, having been uncovered by the action of the winds.

Returning to the objects of shell of purely ornamental character, we would mention flat discs with a central hole, which were probably not strung like the ordinary beads, but arranged in some other way. Quite a number of these discs, made of fluviatile shells, were found some years ago in the now leveled "Big Mound" at St. Louis. Some of them measure more than an inch in diameter. The collection contains similar discs perforated with several holes, and sometimes ornamented around the circumference, which were obtained from California (Figs. 270 and 271, Santa Cruz Island). They are cut from the *Haliotis*-shell. Increasing in size, the shell discs assume the character of gorgets, which were worn suspended from the neck, or attached in some way to the dress. They are round or oval plates, from two to four inches and more in diameter, on which designs are engraved or cut through. The ornamentation is traced on the concave side, which formerly exhibited the shining part of the shell. The collection contains, among other specimens, an ornamented shell gorget from Tennessee, which is now perfectly bleached by age, but evidently formed a beautiful decoration while in its original state (Fig. 272). It is pierced with two holes for suspension and with two lateral holes, probably intended for further fastening. The ornamental tracing on this specimen bears a striking resemblance to the pattern engraved on a shell gorget figured on Plate XXX of the "Antiquities of the Southern Indians" by Charles C. Jones. The similarity in the designs of such shell plates has been pointed out by the late Professor Jeffries Wyman.[7] Some shell discs are ornamented with regularly disposed perforations, and others are entirely plain, showing only the holes for suspension (Fig. 273, mound in Kentucky). Such specimens, whitened by having lain for centuries in the ground, offer now little attraction to the eye, though they must have constituted beautiful ornaments when exhibiting the pearly coating of the shell.

The round or oval gorgets just described are made from *Bysicon*-shells, which were also employed in the production of another class of large ornaments, representing very rudely executed human faces. They are pear-shaped, from five to six inches long, and about four inches wide in the broadest part, where they are pierced with two small holes, evidently intended for eyes. A slight elevation marks the nose, below which there is sometimes seen a third hole indicating the mouth. In addition, the surface is often ornamented with incised lines. The decoration in these typical objects, which probably served as gorgets, is executed on the convex part of the shell. They have been found in mounds of Tennessee, and elsewhere.

Shell-matter was wrought into a variety of other ornaments designed for suspension or attachment. In most instances the fastening was effected by perforations, but exceptionally by grooves, as in the case of a small pendant, pear-shaped in outline, which may have formed an appendage to a string of

---

[7] The tracing often shows the figure of a coiled rattle-snake.—Fifth Annual Report of the Trustees of the Peabody Museum, Boston, 1872, p. 17.

beads (Fig. 274, New York). The Southern Californians chiefly employed the nacreous *Haliotis*-shell as the material for their ornaments, which are abundantly represented in the collection of the National Museum. There are, for instance, ornaments shaped like a ring, provided with a pierced stem projecting from its circumference (Fig. 275, Santa Cruz Island). Such objects may have been worn as ear-pendants. Another class of Californian ornaments cut from the *Haliotis*, consists of somewhat crescent-shaped pieces truncated at their extremities, and pierced for suspension (Fig. 276, Dos Pueblos). They may have been worn as gorgets. Still other objects of decoration (?) are cut in a variety of hardly definable, irregular forms, which present, however, generally rounded outlines (Fig. 277, Santa Cruz Island; Fig. 278, Dos Pueblos). The holes drilled through them characterize them as objects designed to be suspended or attached.[1] The Californian specimens here treated, although stained by age, retain much of their iridescent nacre, and a more recent origin must be ascribed to them than to the described shell objects taken from mounds in the eastern portion of the United States. Lastly, there should be noticed among the Californian specimens a peculiar class of relics cut from the shell of *Lucapina crenulata*, and approaching in shape an oval, from which the middle portion has been removed, leaving an oval hole (Fig. 279, San Miguel Island). As yet it is not known whether articles of this description formed ornaments, or were employed in a more profitable manner.

---

[1] Mr. Paul Schumacher figures in the manuscript report of his explorations in Southern California drawings of worked and pierced pieces of shell, somewhat resembling the original of Fig. 278. These objects, he thinks, were fastened to the end of fishing-lines to attract the prey, in accordance with the present mode of trolling with a spoon-hook.

V. CLAY.

In treating of North American manufactures of clay, it appears proper to begin with those of a useful character, that is, with the vessels employed by the aborigines for culinary and other purposes. Before the advent of the whites, pottery formed an important branch of industry among the eastern Indians, who discovered, however, soon after their contact with the whites the superiority of the metallic vessels which they obtained in trafficking with them, and consequently ceased to manufacture pottery at a very early period. On the other hand, many tribes in the Western Territories (New Mexico, Arizona, Utah, etc.) still practise the ceramic art, producing earthenware of a very creditable character, numerous specimens of which are preserved in the collection of the National Museum. On a rough estimate, it may be said that the art of pottery, as practised in the aboriginal fashion, has become extinct in the eastern half of the United States. There is indeed, still some pottery made by Indians in that part of the Union, but it hardly can be called Indian pottery. Thus, the Catawba Indians, residing upon the banks of the Catawba River in York County, South Carolina,—an insignificant remnant of a once powerful tribe—still make a kind of unglazed pottery, not according to aboriginal taste, but in close imitation of the ceramic productions of the whites. Instead of bowls and cooking-pots of the Indian type, they manufacture cups and saucers, tea-pots, pitchers and basins, flower-pots, and other species of earthenware of patterns altogether distinct from the models in vogue among their forefathers. The writings of early and even comparatively modern authors on North America are not deficient in particulars relating to the art of pottery among the natives occupying the eastern area of the present United States. According to their statements, those tribes were most advanced in the manufacture of earthenware, who inhabited the large tracts of land formerly called Florida and Louisiana, which comprise at present the Gulf States and those adjacent to the Lower Mississippi, and their testimony is fully corroborated by the character of such specimens of pottery from those parts as have escaped destruction and are preserved in the collections of the country. Though the sites of ancient Indian settlements are frequently strewn with innumerable fragments of pottery, entire vessels of early date have almost exclusively been obtained from mounds and other burial-places, where they had been deposited by the side of the dead, either for holding food, or designed to be of service to the deceased in his fancied world of spirits.

The manufacture and character of Indian pottery have been described by Du Pratz, Dumont, Adair, Loskiel, and various other authors. "The women," says Du Pratz, in treating of the pottery of the natives of Louisiana, "make pots of an extraordinary size, jars with a small opening, bowls, two-pint bottles with long necks, pots or jugs for preserving bear oil, holding as much as forty pints, and, finally, plates and dishes in the French fashion."[1] Dumont, who likewise describes the manners of the Indians of Louisiana, has left a more minute account of the method they employed in making earthenware. He says: "After having amassed the proper kind of clay and carefully cleaned it, the Indian women take shells which they pound and reduce to a fine powder; they mix this powder with the clay, and having poured some water on the mass, they knead it with their hands and feet, and make it into a paste, of which they form rolls six or seven feet long and of a thickness suitable to their purpose. If they intend to fashion a plate or a vase, they take hold of one of these rolls by the end, and fixing here with the thumb of the left hand the centre of the vessel they are about to make, they turn the roll with astonishing quickness around the centre, describing a spiral line; now and then they dip their fingers in the water and smooth with the right hand the inner and outer surface of the vase they intend to fashion, which would become ruffled or undulated without that manipulation. In this manner they make all sorts of earthen vessels, plates, dishes, bowls, pots, and jars, some of which hold from forty to fifty pints. The burning of this pottery does not cause them much trouble. Having dried it in the shade, they kindle a large fire, and when they have a sufficient quantity of embers, they clean a space in the middle, where they deposit their vessels and cover them with charcoal. Thus they bake their earthenware, which can now be exposed to the fire, and possesses as much durability as ours. Its solidity is doubtless to be attributed to the pulverized shells which the women mix with the clay."[2] Adair, more than a century ago a trader with the tribes who occupied the southern portion of the present Union, states as follows: "They make earthen pots of very different sizes, so as to contain from two to ten gallons; large pitchers to carry water; bowls, dishes, platters, basins, and a prodigious number of other vessels of such antiquated forms as would be tedious to describe and impossible to name. Their method of glazing them is, they place them over a large fire of smoky pitch-pine, which makes them smooth, black, and firm. Their lands abound with proper clay for that use."[3] A very good account relating to the art of pottery, as formerly practised by the tribes of the Mississippi Valley, is given by Hunter: "In manufacturing their pottery for cooking and domestic purposes," he says, "they collect tough clay, beat it into powder, temper it with water, and then spread it over blocks of wood, which have been formed into shapes to suit their convenience or fancy. When sufficiently dried, they are removed

---

[1] Du Pratz: *Histoire de la Louisiane*, Paris, 1758, Vol. II, p. 179.
[2] Dumont: *Mémoires Historiques sur la Louisiane*, Paris, 1753, Vol. II, p. 271.
[3] Adair: History of the American Indians, London, 1775, p. 424.

from the moulds, placed in proper situations, and burned to a hardness suitable to their intended uses. Another method practised by them is, to coat the inner surface of baskets, made of rushes or willows, with clay, to any required thickness, and when dry, to burn them as above described. In this way they construct large, handsome, and tolerably durable ware; though latterly, with such tribes as have much intercourse with the whites, it is not much used, because of the substitution of cast-iron ware in its stead. When these vessels are large, as is the case for the manufacture of sugar, they are suspended by grape-vines, which, wherever exposed to the fire, are constantly kept covered with moist clay. Sometimes, however, the rims are made strong, and project a little inwardly quite round the vessel so as to admit of their being sustained by flattened pieces of wood slid underneath these projections and extending across their centres."[4]

It would be erroneous to suppose the art of manufacturing clay vessels had been in use among all the tribes spread over this widely extended country; for, though exhibiting much general similarity in character and habits, they differed considerably in their attainments in the mechanical arts. Some of the North American tribes, who did not understand the fabrication of earthen vessels, were in the habit of cooking their meat in water set to boiling by means of heated stones which they put into it, the receptacles used in this operation being large wooden bowls or troughs, water-tight baskets, or even the hides of animals they had killed. The Assineboins, for example, cooked in skins, as described by Catlin.

Generally speaking, the aborigines of North America acquainted with the art of pottery formed their vessels by hand, modeling them sometimes in woven baskets of rushes or willows, and were, as far as we know, unacquainted with the art of glazing. They mixed the clay used in their pottery either with pounded shells or sand, or with pulverized silicious rocks; mica also formed sometimes a part of the composition. In many cases, however, the clay was employed in an unmixed state. Their vessels were often painted with ochre, producing various shades, from a light yellow to a dark brown, or with a black color. They decorated their pottery with incised straight or curved lines or combinations of lines and dots, and embellished it also by notching the rims, or surrounding them on the outside with studs or in various other ways. The vessels exhibited a great variety of forms and sizes, and many of them had rounded or convex bottoms. The aborigines hardened their earthenware in open fires or in kilns, and, notwithstanding the favorable statements of some authors, it was much inferior in compactness to the common ware manufactured in Europe or America.

These remarks, it should be understood, apply to the pottery made by the Indians who inhabited the eastern half of the United States. A superior kind

---

[4] Hunter: Manners and Customs of several Indian Tribes located west of the Mississippi, Philadelphia, 1823, p. 296.

of pottery was manufactured in the more western regions of the continent, as shown by numerous fragments of ancient earthenware which occur, for instance, on the Little Colorado and Gila, especially among ruins, and are often highly decorated and painted with various colors, exhibiting a style of workmanship differing from, and surpassing that, which prevailed on the eastern side of the Rocky Mountains. The superiority of *Mexican* pottery compared with that of the more northern tribes is too well known to be particularly dwelled upon.

The simplest clay vessels left by the eastern aborigines are bowls of a more or less semi-globular shape, cut off abruptly at the rim and destitute of decoration or any kind of handles. Such specimens vary much in size, and are often of rude workmanship. The more elaborate articles of this class, however, show two or more projections immediately below the rim. Of this class is a vessel with four small horizontal projections, probably put on for the sake of convenience as well as for ornament. In this specimen the clay is mixed with particles of coarsely pulverized shells (Fig. 280, mound in Tennessee). This vessel is not of very good workmanship. Much better is a round bowl of larger size, provided on one side with a handle in the shape of the head and neck of a bird (perhaps intended for a duck), and balanced, as it were, on the other by a plain handle rising obliquely from the rim. With some imagination the bird's tail might be recognized in the second handle (Fig. 281, mound in Illinois). In this specimen the clay is slightly mixed with pulverized shells, and the outside was originally painted brown. Similar bird-shaped bowls have been figured and described; also such in which a human head takes the place of that of a bird. Bowls of a more elaborate shape contract more or less toward the aperture, where they terminate in a rising rim. Such bowls are often furnished with projections or ears for facilitating handling. A specimen of this kind (Fig. 282), which was taken from a mound in Union County, Kentucky, is set with four ears around the circumference. Another bowl, formed of clay strongly mixed with pounded shells, shows four equidistant small projections in the plane of the aperture. The shoulder portion is ornamented with crescent-shaped impressions (Fig. 283, mound in Tennessee). A third specimen of the class under consideration is furnished with two mutilated studs projecting below the shoulder (Fig. 284, Arkansas). It is shaped with tolerable regularity and much better burned than any of those thus far described. This vessel seems to have been originally coated with black paint. Small particles of shells are visible in the clay.

A peculiar, though by no means uncommon type is shown in a fine specimen very broad near the bottom and contracting, without forming a shoulder, toward the comparatively narrow aperture. This vessel (Fig. 285, mound in North Carolina) is flat-bottomed and ornamented on the outside with deeply incised curved lines, distributed in regular patterns. There are small particles of mica and of other stone perceivable in the mass of the clay.

Vessels in which the portion projecting above the shoulder becomes narrow and forms a kind of neck approach the bottle shape. Of this character is a well-made and elegantly formed specimen from a mound in Tennessee (Fig. 286). The original paint, a bright red, has not been totally effaced by time. A somewhat smaller, but very gracefully shaped vessel of this kind, which is

CLAY VESSELS (¼).

ornamented with regular figures formed by circles and other curved lines radiating from them, was discovered in a Louisiana mound (Fig. 287). This vessel narrows toward the flat bottom, and its cylindrical neck is provided with a prominent lip. It appears to consist of pure or nearly pure clay, is of a light-brown color passing into black in some places, and has hardly suffered

78 POTTERY.

from the effects of time. Vessels of this description, though resembling each other in general contour, present a great variety of shapes, but they are in most cases less carefully moulded than the two specimens just described. Some are small, measuring only a few inches in height. A specimen from a mound in Tennessee (Fig. 288), by no means the smallest in the collection, is

CLAY VESSELS (⅓).

four inches and a half high, and consists of unpainted clay, with the usual admixture of triturated shells. A larger vessel with a wide neck is distinguished by a rather tasteful ornamentation and a reddish brown paint still adhering to the clay (Fig. 289, grave near Milledgeville, Georgia). These vessels with high and wide necks may be considered as typical. Of a quite different shape is a flat-bottomed ornamented specimen inwardly curved towards the bottom, and provided with a narrow mutilated neck (Fig 290, mound in Louisiana).

CLAY VESSELS (¼).

The collection contains a number of large vessels which, on account of their long and narrow necks, present the true bottle shape. A well-preserved specimen of this kind (Fig. 291) was obtained from a Tennessee mound. The neck slightly expands at the aperture, and where it joins the body of the vessel it is surrounded by eight ornamental studs set in pairs. This vessel was never painted, and therefore shows the natural gray color of the clay, in which numerous diminutive fragments of shell can be seen. One of the finest pieces of pottery in the collection (Fig. 292) is a bottle-shaped jar furnished with a stout and convenient handle. The mutilated neck only shows a somewhat rude linear ornamentation. This specimen, which consists of a gray unpainted clay, mixed with small particles of a black mineral substance, was taken from a mound near Provo, Utah Territory.

There are in the collection some very large vessels which undoubtedly were designed for cooking purposes. One of them (Fig. 293) is more than fourteen inches high, and measures nearly thirteen inches across the aperture. The portion below the rim shows a depression which rendered suspension practicable. This method had to be resorted to, because the kettle could not stand on its lower part which presents an almost conical shape. The outer surface of the vessel shows impressions of tolerably regular pattern and apparently not traced by hand, a circumstance rendering it probable that the vessel was modeled in a woven basket. This remarkable specimen was ploughed up not far from Milledgeville, Georgia. Large clay vessels of a more elongated form,

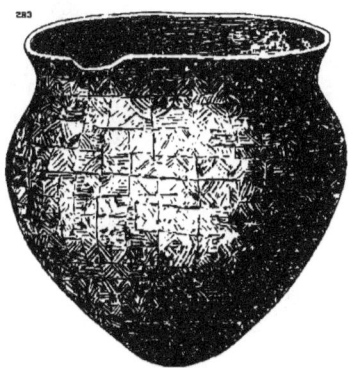

CLAY VESSEL (⅓).

though less conical at the bottom, undoubtedly were employed as funeral vases among certain tribes of the South, for several such vessels containing human bones have been taken from southern mounds. A specimen of this description is preserved in the National Museum. This vase, which was badly injured during its exhumation, resembles in general outline and size the specimen just described. The depression below the rim is somewhat shallower, the lower portion more rounded, and the outside shows impressions of a rather indistinct character. The vessel was discovered in a low mound on the Oconee River, nine miles below Milledgeville, Georgia. When found, it was covered with a well-fitting arched lid, and contained unburned human bones, which soon crumbled to dust upon exposure to the air (Nr. 12305 of the collection).

The largest vessels made by the Indians, it seems, were those used in procuring salt by evaporation near salt springs. In such localities there have been found thick fragments of rude earthenware, bespeaking vessels as large as a barrel. This kind of pottery is usually mixed with coarsely pounded shells. The collection contains such fragments derived from Tennessee and other States, but no entire or nearly entire vessel, and the writer is not aware that a perfect specimen is preserved in any collection of the United States.[5]

Among the large number of smaller vessels in the Museum we take notice of one which is remarkable for a depression encircling its middle, giving the object almost the appearance of two bowls, one placed upon the other (Fig. 294, mound in Louisiana). This specimen, which is flat-bottomed and rudely ornamented with lines and dots, represents a type, though not one that is very frequently met. Similar vessels are still made by the Zuñi Indians.[6] A very

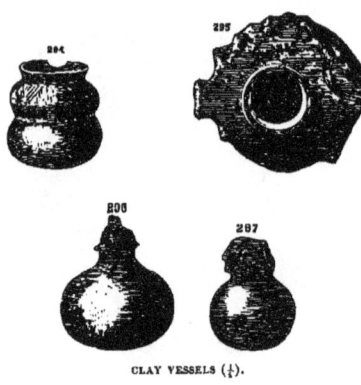

CLAY VESSELS (¼).

curious vessel, obtained from a mound in Tennessee, is made in imitation of a fish, in which ichthyologists have recognized the sun-fish (*Pomotis*), an inhabitant of the Mississippi River. The neck, about an inch in height, rises from the right side of the fish (Fig. 295, upper view). A smaller and less elaborate specimen of the same shape, taken from a mound in Louisiana, is preserved in the Museum. Such forms might be looked for in collections of ancient Peruvian pottery.

Lastly, mention must be made of a class of vessels which bear some resemblance to bottles, in which the upper part or neck forms the imitation of a human head, or of that of an animal, the aperture being usually placed at the back part of the head. In a vessel of this description, taken from a mound in Tennessee, the upper part bears a slight resemblance to the head of some animal (Fig. 296). Another specimen of this character

---

[5] Since the above was written, there has been temporarily deposited in the Museum by the administration of the Louisville Public Library, a vessel of this description, found in a fragmentary state, but restored so as to show its original form. The specimen in question has the shape of a pan with slightly flaring sides and thickened rim. It measures twenty-six inches in diameter at the rim, and is eight inches deep. The thickness of the bottom and sides does not exceed half an inch. The outside of this vessel shows the impressions of the basket in which it was formed, while the inside is perfectly smooth. The clay is of a grayish color, and mixed with pounded shells.

[6] The natives of British Guiana manufacture pottery of the same form, as shown by several specimens in the collection. The double gourd, it appears, served as the model.

exhibits a human head, with the nose, chin, and ears distinctly marked. The occipital portion forms the aperture (Fig. 297, mound in Union County, Kentucky). Vessels in the shape of rude human figures or of animals occur not unfrequently in the tumuli of the Mississippi Valley.

There are in the collection numerous fragments of pottery from all States and Territories of the Union, and from other parts of North America. Many are large enough to show the original shape of the vessel to which they belonged, while others serve to illustrate the different styles of ornamentation in vogue among the aboriginal potters. Of particular interest are the fragments of pottery obtained among the ruins of ancient settlements on the Little Colorado and Gila, and from other parts of the Western Territories. The specimens, for instance, collected during Lieutenant Whipple's survey of those districts are all in the collection, together with many other interesting objects obtained by his party.[7] The sherds in question betoken a much higher state of the potter's art than that ever attained by the aborigines of that part of the United States which lies east of the Rocky Mountains, and the tribes inhabiting now the localities where such fragments occur, produce no earthenware of equal quality. The fractures of such sherds usually exhibit a compact clay of a gray, yellowish, or light-red color, and they are coated, sometimes on both sides, with durable whitish-gray, yellow, or bright-red paint, forming a ground on which parallel lines, lozenges, and other (sometimes very complicated) patterns are executed in black or in other colors. In many instances the paint on these fragments appears as a thin layer which presents a glossy surface, and is almost as hard as the glaze on the clay vessels made by whites. A number of specimens, however, exhibit no paint, but ornamentation of another character, in the shape of raised or indented figures, which betoken, in many instances, considerable taste and knowledge of the art of decoration. "It may not occur to every one," says Thomas Ewbank in Lieutenant Whipple's report, "that most, if not all, the elements of decorative art, as regards curved and straight lines, which are supposed to have originally occurred to the Egyptians, Assyrians, Greeks, and other advanced nations of the eastern hemisphere, have been exhibited by the ancient occupants of the western one. In the relic just noticed,[8] we have the line rolled spirally inward and outward—the involute and evolute. In other samples of pottery the *guilloche*, or curved fillet, in various forms, is met with; also, waving lines, arched, invected, engrailed, radiant, embattled; the trefoil, cross, scroll, and numerous other initial forms, though less expanded and diversified than in the Old World." Generally speaking, broken pottery answering more or less the description here given, has been found in the Territories of Arizona, New Mexico, Utah and Colorado.

---

[7] Described and figured in Vol. III of the "Reports of Explorations and Surveys to ascertain the most practicable and economical Route for a Railroad from the Mississippi River to the Pacific Ocean." Washington, 1856.
[8] Fig. 12 on page 49 of Whipple's report.

It would be impossible to mention here the numerous specimens of pottery derived from Mexico, where, as every one knows, the aboriginal ceramic art had attained a far higher degree of perfection than in the districts lying northward of the Aztec empire. Attention must be drawn, however, to two large vases of exquisite workmanship, which were brought to the United States by General Alfred Gibbs, after the termination of the Mexican war, and presented, with many other valuable Mexican relics, to the National Museum by his mother, the late Mrs. Gibbs, of New York. One of them (Fig. 298), a most elaborate specimen of pottery, is a round vase standing on four curved feet, and narrowing toward the aperture, which is formed by a short neck terminating in a horizontally projecting rim, ornamented with incised ring-shaped patterns. The vase, which measures thirteen inches and

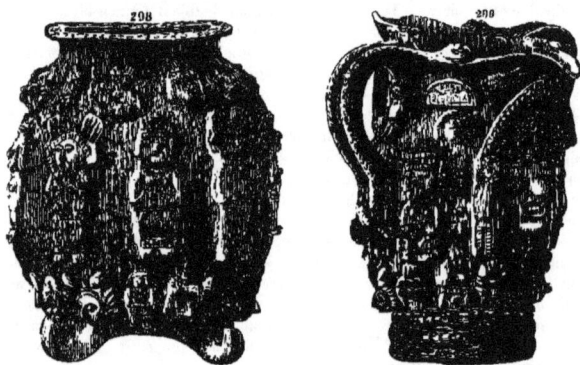

MEXICAN VASES (⅓).

a half in height, is surrounded by ten figures in relief, representing females, five of which grasp a child with the left arm. These five figures, which alternate with the others not holding children, are distinguished from them by a more conspicuous head-dress. Above these ten figures are to be seen, in a row, nine human heads, or masks, and between them, at nearly equal distances from each other, three lizard-like designs, constituting twelve figures in all. The feet divide the lower portion in four compartments, each of which exhibits a figure of a man flanked by two human heads, and each foot is surmounted by an animal, probably representing the coyote, of which only the head, chest, and fore-paws are visible.

The other vase (Fig. 299), matching the one just described, is a still more admirable specimen of Mexican pottery, and, as far as the general outline is concerned, might readily be taken for a vessel of Etruscan or Greek origin. The peculiar ornamentation, however, stamps it at once as a Mexican product of art. The vessel may be compared to a pitcher with two handles standing opposite each other, and with two mouths projecting between them. The handles divide the vase into two halves ornamented nearly alike. Each handle is formed by two snakes, crossing their tails and resting their heads on the rim, and the flat base of the vessel is moulded in the shape of a coiled serpent. A large human head with a prominent chin and protruding tongue, wearing a curious crescent-shaped head-dress with long lateral appendages, constitutes the central figure of each side of the vase. To the right and left of this large head are lizard-like designs, and next to them near the handles, figures of women. On each side of the head-dress surmounting the large head appear three figures, one lizard-shaped, the other in the form of a human head, and the third in that of a crescent-shaped tablet bearing, it has been thought, hieroglyphic signs. The four tablets, it should be stated, exhibit the same characters (Fig. 300, natural size). The lizard-like figure also appears below the termination of each mouth. Beneath each handle the vase bears the moulding of a male figure, and the outer curve of the handles shows, between the bodies of the snakes, the relief design of a fish. The lizard-like figure is seen again on each side of the rim between the serpents' heads.

TABLET (½).

The circumference of the vase exhibits, immediately above the coiled snake forming its base, ten human heads wearing elaborate head-dresses (like all imitations of the human head on this vase), and an eleventh figure of indistinct character, perhaps intended for a hieroglyphical tablet. This remarkable vase is fourteen inches and a quarter in height, and coated with black paint, like the specimen previously described.

Another beautiful Mexican vase of somewhat globular shape (Fig. 301) is remarkable for its elaborate raised ornamentation, which consists of four entwined snakes and four masks placed at equal distances from each other. The vessel stands on three feet presenting beautifully executed eagles' heads. The color of the vase is a light reddish-brown.

There are in the collection many small Mexican vessels, a full description of which would exceed the limits of this account. We only notice among them a small vessel tapering to a point at the lower extremity, reminding one of similar productions of ancient Roman art (Fig. 302, Tezcuco), and to a well-shaped goblet of red ware, derived from Sacrificios Island, nearly opposite Vera Cruz (Fig. 303).

Having treated of North American clay vessels, we have to notice the fabrics of clay not intended for culinary and other domestic uses. The North American Indians frequently made their pipes of clay, moulding them into various forms, sometimes with great ingenuity, as previously stated. They also manufactured clay images, which have been noticed in different publica-

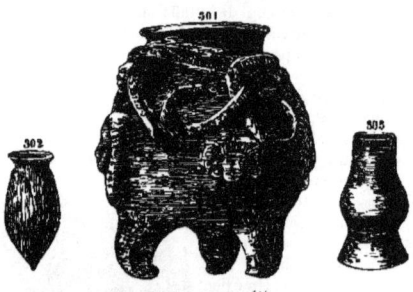

MEXICAN VESSELS (¼).

tions. Generally speaking, such imitations of the human form are of a primitive and uncouth character, and inferior to corresponding manufactures of stone. Much better than the ordinary aboriginal clay fabrics of this kind is a head which, to judge from the remaining part below the neck, may have originally formed the handle of a vessel. In this head the features are clearly, though not correctly, defined (Fig. 304). The head is hollow and pierced

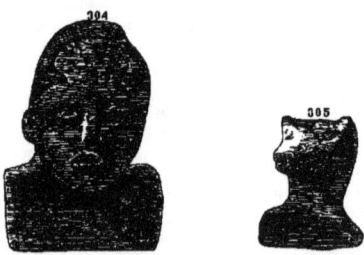

TERRA-COTTA FIGURES (¼).

at the occiput with a hole, which evidently has been enlarged after the discovery of the relic. This specimen was found among shell-heaps near Mobile, Alabama. The same locality has furnished a rude aboriginal clay manufacture in the shape of a wolf's head, to all appearance likewise the handle of a vessel (Fig. 305). In this instance the specimen is solid, consisting of clay with the usual admixture of shells.

The ancient Mexicans, on the other hand, have left numberless clay figures representing the human form, which are, however, generally more conspicuous for elaborate details than for correctness of the proportions, the heads being often unnaturally large. The significance of many of these figures is not known, though it may be assumed that a large proportion of them relates to the mythology of the Aztecs. Some may represent household gods, or penates, while others, perhaps, were nothing else but toys. Most of these manufactures are hollow and pierced with a few holes for emitting the heated air produced by the baking. Without this precaution the objects would have burst, owing to the expansive force of the air.

One of the most elaborate Mexican figures of the collection (Fig. 306) represents a man seated, with the hands resting on the knees, and bearing on his back another human figure so placed that its head surmounts that of the first, while its hands press against the forehead and its feet rest on the shoulders of the lower figure. The upper figure wears a rather low head-dress, and the lower one is profusely decorated with armlets, wrist-bands and leg-ornaments. The most conspicuous attributes of this curious pair consist in two serpents which, descending from the head-dress of the upper figure, encompass, as it were, the group on both sides, and rest their heads between the feet of the lower figure. In this specimen the clay is well burned and shows externally a light-brown paint.

A Mexican image of simpler design (Fig. 307), likewise represents a man in the attitude so often exhibited in Mexican and Central American terracottas and sculptures, namely, seated and placing the hands on the knees. The figure is highly ornamented and wears a head-dress of a shape reminding one of a terraced pyramid. The color is a pale red.

Two remarkable figures of the collection, nearly identical in shape, though somewhat differing in size, were presented to the National Museum by the family of the late George Gibbs. They are of a more uncouth appearance than the two specimens before described, and represent squatting women pressing their hands against the ears (Fig. 308). The faces indicate aged individuals with prominent noses and somewhat protruding tongues. The sexual parts are broadly marked. The peculiar head-dresses show, in both instances, on the right side a projection resembling a tuft of feathers. Both figures are coated with a shining black color. It would be interesting to know the circumstances which gave rise to the manufacture of these two almost identical images. Quite different in design is a small statuette of a woman dressed in an ornamented gown reaching to the feet, and wearing a high cap (Fig. 309). The hollow figure encloses a loose clay ball, giving the object the character of a rattle. Rattles of clay, it is well known, belong to the common relics of the ancient Mexicans.

86   POTTERY.

The peculiar attention paid to snakes by the inhabitants of Anahuac is exemplified in the collection by a number of mouldings in clay representing

MEXICAN STATUETTES.

those reptiles in various attitudes. Several of those specimens show a snake coiled on the back of a turtle and in the act of biting its head. In some of

these representations the lower part of the turtle's neck exhibits a human
face (Fig. 310). This curious group
is quite typical, and probably refers to
some tradition or to a religious conception of the Aztecs. A coiled snake
with uplifted head is likewise frequently
met among Mexican terra-cottas, and a
number of productions of this character
can be seen in the National Museum.
One of them (Fig. 311) is the well-executed figure of a rattle-snake with
four rattles. Such specimens are usually solid, exhibiting externally a shining
black or other dark color.

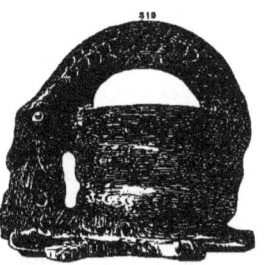

Clay was employed by the Mexicans
in the manufacture of small mask-shaped
heads and of various other objects, either of a useful or ornamental character.
Their whistles and rattles exhibit an endless variety of forms, being made in imitation of the human figure, or in the
shape of animals, or representing monstrous creations of fancy which it would
be difficult to define. Sometimes the feet
of Mexican vessels were made hollow
for receiving clay balls, insomuch that
such objects partook of the combined

COILED SNAKES IN TERRA-COTTA (⅓).

characters of utensils and of toys. The Mexican clay spindle-whorls (mala-

MEXICAN SPINDLE-WHORLS (½).

catl), of a nearly semi-globular shape, are often tastefully ornamented, as
shown by several specimens among the Mexican relics in the Museum (Figs.
312 and 313, Tezcuco).

VI. WOOD.

Among the materials composing North American aboriginal relics we assign the last place to wood, considering that the occurrence of wooden manufactures of early date is extremely limited. A substance so much subject to decay as wood cannot be expected to resist physical influences for a considerable length of time, unless peculiar circumstances retard its destruction. Thus the ancient Swiss lake-villages have yielded an abundance of wooden articles, owing to the preservative qualities of the peat enclosing them, which had accumulated along the lake-shores. The National Museum contains but a small number of wooden objects which can be included in the archæological series, and these were almost exclusively obtained from graves of the Californian Santa Barbara Islands.[1] The articles apparently consist of cedar wood, which has become very light, almost as light as the wood of the utensils extracted from the sites of lacustrine settlements in Switzerland. Among these Californian relics are rotten wooden handles, some, indeed, still holding arrow-head-shaped knife-blades of flint, cemented into the wood by means of asphaltum. They resemble the Pai-Ute knives mentioned in the beginning of this account (page 2). There is further to be noticed a wooden bailing-vessel with a short handle, fitting in a rectangular hole cut into the vessel (Fig. 314, Santa Cruz Island). A number of well-made toy canoes, the smallest of which measures seven inches in length, bears witness to the maritime propen-

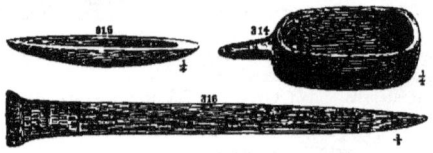

OBJECTS OF WOOD.

sities of those islanders (Fig. 315, Santa Cruz Island). These specimens are very interesting, as they undoubtedly represent the shape of the "dug-outs" used by the Southern Californians. It is known, however, that they also em-

---

[1] The writer is at this moment unable to state whether these relics were found associated with manufactures of Caucasian origin or not.

ployed boats constructed of planks. Perhaps the most curious wooden object from California is an implement resembling a short sword, terminating opposite the point in a broad flat handle, inlaid with a kind of mosaic of *Haliotis*-shell (Fig. 316, Santa Cruz Island). It is stated that "sabres of hard wood with edges that cut like steel" were among the weapons of the California Indians (H. H. Bancroft); but the object in question is neither sharp-edged, nor, as it appears, made of very hard wood, and, being, moreover, thin and light, hardly could have formed an efficient weapon. Hence there is a probability that it represents either a weapon of parade, or some kind of implement designed for peaceable purposes. From the same localities were derived parts of planks and other fragmentary articles of wood, the use of which cannot now be determined.

By far the most remarkable relic of vegetable substance in the collection is a piece of matting of split cane, fifteen inches long and about nine inches wide, which was found under very peculiar circumstances on Petite Anse Island, near Vermilion Bay, on the coast of Louisiana. A notice by Professor Henry, affixed to the specimen in question, runs thus: "Petite Anse Island is the locality of the remarkable mine of rock salt, discovered during the civil war, and from which, for a considerable period of time, the Southern States derived a great part of their supply of this article. The salt is almost chemically pure, and apparently inexhaustible in quantity, occurring in every part of the island (which is almost five thousand acres in extent) at a depth below the surface of the soil of fifteen or twenty feet. The fragment of matting was found near the surface of the salt, and about two feet above it were remains of tusks and bones of a fossil elephant. The peculiar interest in regard to the specimen is in its occurrence *in situ* two feet below the elephant remains, and about fourteen feet below the surface of the soil, thus showing the existence of man on the island prior to the deposit in the soil of the fossil elephant. The material consists of the outer bark of the common southern cane (*Arundinaria macrosperma*), and has been preserved for so long a period both by its silicious character and the strongly saline condition of the soil." This specimen, so interesting on account of its associations, was presented to the National Museum by Mr. J. F. Cleu.

SUPPLEMENT.

Since the preceding pages were written, the National Museum has been enriched with a large number of aboriginal relics, some of them belonging to

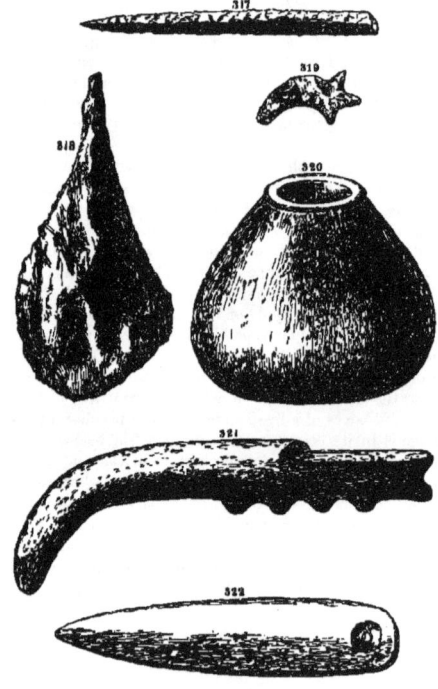

STONE IMPLEMENTS (¼).

types never before described. It would be impossible to notice at present the extensive additions to the collection, but a few typical objects which appeared of particular interest to the writer may here be mentioned.

Fig. 317.—A well-wrought three-sided perforator of brown flint, obtained with other tools of the same description from Santa Cruz Island, California.

Fig. 318.—Another piercing tool of large size and consisting of light-gray flint. It terminates in a three-sided point which is rounded by wear. The portion opposite the point is broad and massive. In other specimens of this class, all of which were found by Mr. Paul Schumacher on Santa Cruz Island, the thick part is coated with asphaltum, doubtless for more convenient handling.

Fig. 319.—A chipped sickle-shaped implement of light-gray hornstone, probably used in scraping round objects of wood, bone, etc., the inner curve forming a strong, carefully wrought edge. The specimen, which was found in Ohio, terminates in an indented tang or stem, by which it appears to have been attached to a handle.

Fig. 320.—A remarkable specimen of the perforated club-head-shaped articles previously described. It consists of greenstone, and was found in California, like the other objects of analogous form before noticed.

Fig. 321.—A very singular tool made of dark basaltic rock. The working part is curved, and the upper side shows a shoulder on which the end of a handle may have rested. The four conical elevations seen on the lower side appear to have served for confining the ligatures by which the handle was connected with the implement. This specimen was obtained in Oregon.

Fig. 322.—One of several objects made of argillite and obtained from Massachusetts. They are flattish and about one-third of an inch thick at the perforated rounded end, but become gradually thinner toward the tapering opposite extremity. It may be assumed that they served as implements, though their special use thus far has not been ascertained. In some the perforation is wanting, which hardly would be the case if these objects had been designed for other than useful purposes.

Among the objects lately obtained from Utah is a large stone of somewhat compressed roundish form, and showing no other modification but a groove running across the broader sides. The material of this stone, which weighs fifteen pounds, is vesicular basalt. The writer has seen in the State of New York, and elsewhere, similar stones or boulders grooved in the same manner. They are thought to have served instead of anchors.

APPENDIX 1.

The Aboriginal Modes of hafting Stone and Bone Implements.

Various North American tribes still use, though to a limited extent, weapons and tools of stone and bone, hafting them according to the methods in vogue among their forefathers. Such modern specimens illustrate the manner in which the stone axes, celts, adzes, and other implements of earlier date were rendered serviceable by the addition of handles, and it has been thought proper, therefore, to figure and describe here the most characteristic among the numerous hafted weapons and tools preserved in the ethnological department of the National Museum.

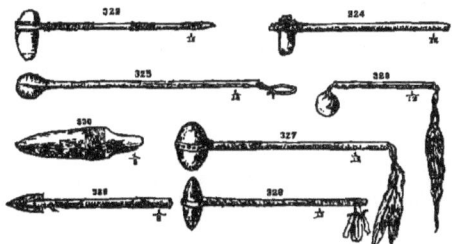

HAFTED STONE WEAPONS.

Fig. 323.—Grooved greenstone axe with a hickory withe bent around the groove. The ends of the withe, which form the handle, are firmly bound with strips of raw-hide below the stone head, near the middle, and at the lower part (Dakota Indians).

Fig. 324.—Polished celt of argillite, chipped thin at the blunt part to fit into the cleft end of an oaken stick, where it is secured by twisted cords of sinew (Indians of the Missouri Valley).

Fig. 325.—War-club, consisting of a heavy roundish stone firmly connected with a long handle. Both the stone and the handle are tightly cased in raw-hide sewed together with sinew. The end of the handle is perforated for receiving a loop of dressed skin, designed to pass around the wrist (Dakota Indians).

(93)

APPENDIX.

Fig. 326.— A weapon of similar character. In this instance, however, the handle is much shorter, and the round stone head is not firmly attached to its end, but is merely connected with it by flexible thongs. The raw-hide covering of the weapon (including head and handle) consists of one piece taken from the caudal portion of an ox, a part of whose tail forms an ornamental appendage to the handle (Apaches). The analogy of such weapons to the mediæval "morning-stars" has been pointed out on page 32.

Fig. 327.— A war-club with a well-wrought and polished egg-shaped head of yellowish limestone, grooved around the middle for receiving the handle. One end of the latter is bent like a hoop to fit into the cavity of the stone, and strengthened by a casing of raw-hide, which extends about six inches below the head. The part of the ashen handle that encircles the stone is ornamented with large-headed brass nails. The extremity of the handle, again, is enveloped by a tightly fitting covering of raw-hide, taken from the caudal part of a buffalo. A tuft of the animal's tail has been retained for the sake of decoration, and a feather of the wild turkey is attached to the hair by means of a narrow strip of dressed skin (Blackfeet).

Fig. 328.— A weapon of the same description. The polished head, which consists of greenstone, is smaller and more elongated than in the original of Fig. 327. The handle shows the usual casing of raw-hide, and is pierced at the lower extremity for facilitating the attachment of a wrist-strap (Missouri River Valley).

Fig. 329.— Dagger-knife, chiefly used as a hunting weapon. It consists of a ground lance-head-shaped blade of dark slate, inserted and riveted by means of a wooden peg into a barbed ivory socket, which is attached to a short cylindrical handle of pine-wood (Natives of Nunivak Island, Alaska).

Fig. 330.— Scabbard of the dagger-knife just described. Formed by two hollowed pieces of pine. which are held together by a binding of split spruce-roots.

Fig. 331.— Grooved hammer of greenstone, the flattened lower side of which rests against a corresponding flat part of the curved handle. The head is connected with the handle by ligatures of raw-hide (Fort Simpson, British Columbia). There are similar hammers from the Northwest Coast in the collection, in which the narrower part of the stone is formed in imitation of an animal's head. Other hammers or mauls cased in raw-hide, one of which has been previously figured and described, are still in use among various tribes (See Fig. 79 on page 20).

Fig. 332.— Large adze-shaped pick of whalebone, attached by raw-hide thongs to a flattish massive pine handle, which is perforated at the broader part for receiving the ligatures. The latter are confined by notches in the sides of the head (Mackenzie's River District).

APPENDIX.   95

Fig. 333.—Smaller implement of the same character. The head of walrus ivory and the short pine handle show corresponding perforations, serving to connect both parts by means of raw-hide ligatures (Nunivak Island).

Fig. 334.—Hoe made of the shoulder-blade of a buffalo, and bound with raw-hide thongs to the shorter flat part of a hook-shaped curved handle of ash-wood. A pad of dressed skin is placed between the blade and the corresponding part of the handle (Arickarees, Fort Berthold, Dakota Territory).

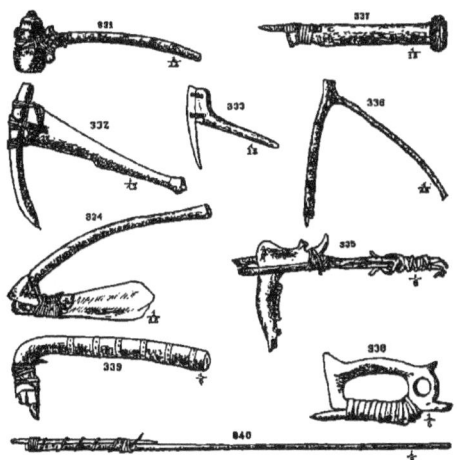

HAFTED STONE AND BONE TOOLS.

Fig. 335.—Implement marked "reaping-hook." It consists of the right lower jaw of an antelope, around which is bent a sapling forming the handle. Its two ends are bound together by a strip of bark. The jaw is further secured to the handle by a thong of raw-hide (Caddoes, Indian Territory).

Fig. 336.—Small celt-shaped adze of argillite, resting against a shoulder at the extremity of a forked handle, the thinner branch of which, being held in the right hand, doubtless served to guide the implement, while the thicker part of the handle was grasped by the left. The stone blade is held in place by a cord of twisted sinew. The tool is said to have been employed in finishing the inside of canoes, thus combining the characters of an adze and a scraper (Natives of Vancouver's Island). Other methods of hafting adzes are exemplified by Figs. 70 and 71 on page 19.

APPENDIX.

Fig. 337.— Long flat celt-like chisel of argillite, attached to a roughly worked cylindrical handle by a thong of twisted raw-hide. The handle is provided with a shoulder against which the stone rests. The tool evidently was used in connection with a mallet, as indicated by the battered upper end, which is, moreover, confined by a ring of twisted spruce-roots (Vancouver's Island).

Fig. 338.— Celt-shaped chisel of argillite, strongly bound with a strip of leather to a carved handle of peculiar form (Vancouver's Island).

Fig. 339.— Chipped flint scraper, partly enveloped in buckskin, and bound by means of a raw-hide thong to a hook-shaped ornamented handle of elk-horn (Mandans).

Fig. 340.— Tool used in chipping stone arrow-points, perforators, etc. It consists of a slender blunt piece of deer-horn, bound with cotton cord to a wooden rod about the thickness of an arrow-shaft (Indians of Nevada Territory).

APPENDIX II.

System adopted in arranging the Smithsonian Collection illustrative of North American Ethnology.*

I. MAN.

Desiccated Bodies.
Skeletons.
Skulls.
Other Parts of Skeletons.

Casts of Indian Heads in plaster, wax, and papier-mâché.
Photographs, Drawings, and Paintings of Aborigines and of Scenes of Aboriginal Life.

II. CULTURE.

(1.) *Aliment, etc.*
  A. Food.
    1. Mineral Food.
      Salt.
      Clay (mixed with food).
    2. Vegetable Food.
      a. Unprepared.
        Roots.
        Bark.
        Buds.
        Flowers.
        Fruits.
        Seeds.
      b. Prepared.
        Sugar.
        Preserved Fruits.
        Meal.
        Mush.
        Bread or Cake.
    3. Animal Food.
      Dried and smoked Meat of Mammals, Birds and Reptiles.
      Dried and smoked Fish.
      Dried Fish-eggs.
      Roasted and dried Insects and Worms.

  B. Drink.
    1. Decoctions.
      Teas, etc.
    2. Fermented Drinks.
      Cider, Wine and Liquor.
  C. Narcotics.
    Tobacco and its Substitutes.
  D. Medicines.
    1. Mineral Medicines.
      Earths, etc.
    2. Vegetable Medicines.
      Herbs.
      Roots.
      Buds.
      Flowers.
      Seeds.
    3. Animal Medicines.
      Pulverized Bones, etc.
(2.) *Habitations.*
  A. Skin Lodges.
  B. Models of Dwellings.
    Shelters.
    Skin Lodges.
    Yourts.
    Huts (of bark, grass, etc.).
    Wooden Houses.

---

*In this classification Professor O. T. Mason's pamphlet, entitled "Ethnological Directions relative to the Indian Tribes of the United States" (Washington, 1878), has been used to some extent.

APPENDIX.

C. Appurtenances.
  Sweat-houses (models).
  Totem-posts (originals and models).
  Gable-ornaments (carved).
  Locks (wooden).

(3.) *Furniture.*
  Mats (of bark, grass, flax, etc.).
  Screens.
  Hammocks.
  Bed-coverings.
  Head-rests (Hoopa Indians, California).
  Cradles.
  Cradle-boards.
  Chairs.
  Stools.
  Washing-vessels.
  Tubs.
  Pails.
  Boxes.
  Chests.
  Lamps.
  Brooms.
  Fly-brushes.

(4.) *Vessels and other Utensils of Household Use.*
  A. Raw Material.
    Stone.
    Clay.
    Roots.
    Grass.
    Rushes.
    Osiers.
    Splints.
    Wood.
    Horn.
    Skin.
    Membrane.
    Dyes and Cements (for baskets, etc.).
  B. Earthenware.
    Cooking-vessels.
    Ollas.
    Spherical Jars.
    Small-necked Jars.
    Canteens.
    Pitchers.
    Dishes.
    Trays.
    Bowls.
    Cups.
    Ladles.
    Spoons.
    Ornamental Vessels.

C. Carved Horn and Wooden-ware.
    Four-sided Vessels.
    Trays.
    Dishes.
    Bowls.
    Cups.
    Dippers.
    Spoons.
    Ladles.
    Stirring-sticks.
  D. Carved Stone-ware.
    Plates.
    Trays.
    Dishes.
    Bowls.
    Cups.
  E. Water-tight and ordinary Basket-work.
    Cups.
    Bowls.
    Flasks.
    Carrying-bottles.
    Baskets of various forms.
  F. Bark Vessels.
    Trays.
    Bowls.
    Pails.
  G. Gourd Vessels.
    Cups.
    Bowls.
    Carrying-bottles.
  H. Skin and Bladder Bottles.

(5.) *Articles serving in the Use of Narcotics.*
  Pipes.
  Tobacco-pouches.
  Cigar-cases.
  Plates for cutting Tobacco.
  Snuff-grinders.
  Snuff-scrapers.
  Snuff-boxes.
  Snuff-tubes.

(6.) *Receptacles used in Transportation.*
  A. On Foot.
    Pouches.
    Burden-straps.
    Burden-nets.
    Burden-baskets.
  B. With Beasts of Burden.
    Bags.
    Raw-hide Cases.

APPENDIX. 99

(7.) *Clothing.*
  A. Raw Material.
    Fur.
    Raw-hide.
    Wool.
    Hair.
    Vegetable Fibre.
  B. Complete Suits (in part exhibited on lay-figures).
  C. Head-clothing.
    Hats.
    Caps.
    Hoods.
    Head-scarfs.
  D. Body-clothing.
    Robes.
    Blankets.
    Mantles.
    Capes.
    Shirts.
    Tunics.
    Coats.
    Clouts.
    Aprons.
    Skirts.
  E. Hand-clothing.
    Mittens.
    Gloves.
  F. Leg and Foot-clothing.
    Sandals.
    Moccasins.
    Shoes.
    Boots.
    Socks.
    Stockings.
    Leggins.
    Garters.
  G. Parts of Dress.
    Bands.
    Belts.
(8.) *Personal Adornment.*
  A. Head-ornaments.
    Wigs.
    Chignons.
    Hair-pins.
    Tucking-combs.
    Head-bands.
    Feather Head-ornaments.
    Labrets.
    Nose-ornaments.
    Ear-ornaments.
  B. Neck-ornaments.
    Necklaces.
    Neck-bands.
    Collars.
  C. Breast and Body-ornaments.
    Gorgets.
    Ornamental Girdles.
  D. Limb-ornaments.
    Rings.
    Bracelets.
    Armlets.
    Anklets.
  E. Toilet Articles.
    Substitutes for Soap.
    Paints (mostly mineral).
    (Paint-mortars).
    Spatulæ (for face-painting).
    Hair-powder.
    Hair-dye.
    Combs.
    Head-scratchers.
    Tweezers for removing the hair.
    Mirrors.
(9.) *Implements for General Use, for War and the Chase, and for special Crafts and Occupations.*
  A. Implements for General Use.
    1. For Striking.
      Hammers and Mauls.
    2. For Cutting, Sawing, Perforating, etc.
      Knives of various forms.
      Hatchets.
      Adzes.
      Chisels.
      Gouges.
      Wedges.
      Scrapers.
      Skinning Implements.
      Saws.
      Drills.
      Awls.
      Cutting-blocks.
      Tool-boards.
      (Tool-boxes).
      (Whet-stones).
  B. Implements for War and the Chase.
    1. Striking Weapons.
      War-clubs (with or without metallic points or stone weights).
      Tomahawks.

APPENDIX.

2. Throwing Weapons.
   Boomerangs (Moquis, etc.).
   Bolas.
3. Thrusting Weapons.
   Knives.
   Daggers.
   Swords.
   Lances.
4. Projectile Weapons and Appurtenances.
   Arrows.
   Bows.
   Quivers.
   Wrist-guards.
   Harpoons and Throwing-boards.
   Slings.
5. Defensive Weapons.
   Shields.
   Helmets.
   Visors.
   Body-armour.

C. Implements for Special Crafts and Occupations.
   1. Implements for Hunting other than Weapons.
      Snares and Traps.
      Nets.
      Hooks for catching small Animals.
      Decoys.
   2. Implements for Fishing other than Weapons.
      Hooks and Lines.
      Sinkers and Floats.
      Nets.
      Traps.
   3. Implements and Utensils used in Gathering and Manufacturing Food.
      Root-diggers.
      Gathering and Winnowing-trays.
      Mortars and Pestles (of wood and stone).
      Stone Troughs or Slabs with Rubbing-stones.
   4. Agricultural Implements.
      Spads.
      Hoes.
      Rakes.
      Reaping-hooks.

   5. Implements for Fire-making.
      Fire-sticks and Drills.
      Flint with Steel and Pyrites.
      Moss.
      Punk.
      Tinder.
      Slow-matches.
      Fire-nests.
      Fire-bags.
   6. Implements for Arrow-making.
      Chipping-tools.
      Shaft-grinders.
      Shaft-straighteners.
      Glue-sticks.
   7. Implements for making Pottery.
      Paddles.
      Smoothing-stones.
   8. Implements for Twisting, Spinning, Weaving, Sewing and Embroidery.
      Fibre-twisters.
      Spindle-whorls.
      Reels.
      Knitting-needles.
      Looms with Woof-sticks and Shuttles.
      Awls.
      Needles.
      Needle-cases.
   9. Implements for Basket-making.
      Plaiting-tools.
   10. Implements for working Skins.
       Scrapers.
       Skin-softeners.
       Burnishers.
       Crimping-tools.
   11. Implements for Carving.
       Knives.
       Gouges.
   12. Implements for Painting (including Paints).
       Bristles.
       Paint-sticks.
       Brushes.
       Rubbing-stumps.
       (Paints).

(10.) *Means of Locomotion and Transportation.*
   A. By Land.
      1. Traveling on Foot.
         Ice-creepers.
         Snow-shoes.

2. Conveyances, etc.
   Saddles.
   Bridles.
   Halters.
   Stirrups.
   Spurs.
   Foot-mufflers.
   Dog-harnesses.
   Reindeer-harnesses.
   Sleds.
   Sleighs.
B. By Water.
   Balsas.
   Dug-outs.
   Bark Canoes.
   Bull-hide Boats.
   Kayaks.
   Oomiaks.
   Pushing-sticks.
   Paddles.
   Oars.
   Bailing-vessels.
   Spear-rests.

11.) *Games and Pastimes.*
A. Gambling Implements.
   Pairs of Bones and Sticks.
   Bundles of Sticks.
   Discs.
   Dice.
   Ivory Blocks and Catching-sticks.
   Cards.
   Chess.
B. Dancing.
   Plumes.
   Wooden Masks and Head-dresses.
   Buffalo-head Masks.
   Head-shields.
   Hip-ornaments.
   Rattles.
   Batons.
   Spears.
   Scalps.
C. Athletic Exercises.
   Rackets.
   Sticks.
   Poles.
   Balls.
   Rings.
   Boundary-sticks.
D. Children's Sports and Toys.
   Dolls.
   Whirligigs.
   Tops.
   Miscellaneous Toys.

(12.) *Music.*
A. Instruments for Beating and Shaking.
   Drums.
   Sounding-bars.
   Rattles.
   Clappers.
B. Rubbing and Stringed Instruments.
   Notched Sticks.
   Cane Harps.
   Cane Fiddles.
C. Wind Instruments.
   Whistles.
   Fifes.
   Flutes.
   Trumpets.
   Horns.
D. Whizzers.

(13.) *Art.*
A. Pictorial Representations and ornamental Designs on Wood, Bark, Bone, Horn, Ivory, dressed Skin and Leather.
B. Carvings in Stone, Wood, Horn, Bone and Ivory.
C. Embroidery and other ornamental Work with Quills, colored Threads, Hair, Feathers, and Beads.

(14.) *Enumeration, etc.*
   Census-sticks.
   Dunning-sticks.

(15.) *Objects relating to Superstitions.*
   Idols.
   Charms.
   "Medicines."
   Medicine-bags.
   Medicine-boxes.
   Batons.
   Rattles.
   Drums.

(16.) *Objects relating to Funeral Rites and Burials.*
   Mourning-yokes.
   Mourning-bracelets.
   Dead-masks.
   Burial-frames.

INDEX.

Adair, 28, 30, 28, 74.
Adzes, with handles, 19.
Alaska, copper articles from, 62.
Appendix I., 93.
Appendix II., 97.
Archaeological series, 1.
Armlets of copper, 61.
Arrow-heads, mostly small, 2.
   " their abundance. 6.
   " are still manufactured, 8.
   " different shapes of, 9, 10.
Arundinaria macrosperma, 89.
Axes, grooved, 19, 20.

Borgers, 44.
Bancroft, Mr. H. H., 31, 70, 89.
Bartrum. 40.
Beads of copper. 61.
   " shell, 98.
   " stone, 51.
Blackmore Museum, England, 63.
Boat-shaped articles, 21, 22.
Bone and horn. 63.
Bowls of stone, 37.

Calumet-pipes. 48.
Carvings of birds, etc. upon pipe-bowls, 46, 47.
Casts of pipes, 43.
   " remarkable relics, 29.
Catlin, Mr., 28, 78.
Catlinite, or red pipe-stone, 50, 51.
Celts of jade in Mexico, 17.
Celts or Wedges. 17.
Chibouc of the Turks, 51.
Chisels, 16.
Chung-kee, an Indian game, 28.
Church, Prof. A. H., 43.
Clams as food, 70.
Clarke. Mr., 22.
Classification of the archaeological series, 2.
Claws as ornaments, 64.
Clay images. 84, 85.
Clay, manufactures of. 73.
   " vessels of, 77, 78, 79, 80.
Clay pipes, 84.
Clow, Mr. J. F., 80.
Club-head-shaped stones. 31.
Coiled snakes in terra-cotta, 87.
Collections from Porto Rico, 4.
Columellæ of shells, 68.
Contents, V.
Cooking vessels, globular. 37.
Copper, 59.
Copper articles from Alaska, 62.
Copper harpoon-heads, 63.
Copper pipes. 45
Coreal. 44.
Cups, hollowed out from the vertebræ of cetaceans, 63.
Cushing, Mr. F H., 63.
Cutting and sawing implements. 13.
Cutting tools, 21. 23.
   " different shapes of, 14.

Dall. Mr. W H., 43.
Davis, Dr. E. H., 51.
Difficulties in determining the real character of stone implements, 2.
Discoidal stones, 28, 29.
Drilled ceremonial weapons, 23.
Drilling in stone, 54.
Drinking cups made of shells, 68.
Dug-outs, 98.
Dumont, 71.
Du Pratz, 28, 30, 74.

Egg-shaped stones, 32.
Epiphysis, pierced for ornament, 65.
Ethnological series, 1.
Ethnology, North American, 97.
Ewbank, Mr. Thomas, 81.

Fifes of bones, 63.
Fish-hooks of bone, 63.
Flakes of flint, obsidian, etc., of different shapes, 8.
Flint knives, hafted, 9.
Foot-tracks, sculptured, 57.
Foster, Mr. J. W., 59.

Gibbs, Gen. Alfred. 82.
Gibbs, Mr. Geo., 85.
Gibbs, Mrs., 63.
Globular cooking vessels, 37.
Gorgets made of shells, 71, 72.
Gouges and adzes, 18, 19.
Gold in grains, 68.
Greenstone, material for implements, 3.
Grinding and polishing tools, 34, 35.

Hafted stone and bone tools, 26.
Hafted stone weapons, 22.
Hammer-heads, grooved, 21.
Hammer-stones, 22.
Harpoon-heads of bone, 63.
Henry, Prof., 81.
Hunter, Mr., 71, 75.

Ice-chisel from Unalaska, 16.
Implements and ornaments of bone, 61.
   " of copper, 59, 60, 61.
   " of stone
       dagger-shaped, 14, 13.
       leaf-shaped. 15.
       wedge or celt-shaped, 17.
Indian tools for making stone arrow-heads. etc., 29.
Introduction, 1.

Jefferson, President Thomas, 56.
Jones. Col. Chas. C., 30, 38, 41, 71.
Jones. Dr. Joseph, 20.
Just, Robert, 45, 50.

Knight of Elvas, 30.
Kohl, Mr., 44.

Lake Superior, ancient mining on the shores of, 59.

Lapis ollaris, 36.
Lartet and Christy, Messrs., 63.
Latimer. Mr. George, 4.
Lawson, 28.
Lewis and Clarke. Messrs., 28.
Loskiel, 74.

Mason. Prof. O. T., 97.
Matting of split cane, 60.
Metate, Mexican, 40.
Mexican carvings. 50.
   " celts of jade, 17.
   " clay figures, 85.
   " knives, 3
   " pottery, superior, 78.
   " spindle-whorls, 87.
   " statuettes, 80.
   " vases. 82, 84.
Modes of hafting stone and bone implements, 92.
   " manufacturing pottery, 74, 75.
Morgan. Mr. L. H., 30.
Mortars of stone and wood, 38, 39.
Motolinia. 8.
Mullers. 42.
Murray, Mr., 2d.

Necklaces of claws, 84.
Needles of bone. 63.
Neolithic period. 7.
Net-sinkers, 27. 29.
Nilsson, Prof.. 36.
Notches in pendants, probably denoting enumeration, 58.
Nut-stones, 40, 41.

Objects of wood. 88.
Obsidian points for arrows, 2.
Ornamented pestles. 43.
Ornaments, 26, 51, 92.

Pai-Utes, 2, 8.
Paint-mortars, 40, 41.
Paleolithic period, 7.
Pebbles, pierced for ornament, 29.
Pendants and sinkers, 26. 27.
Perforations made by drilling. 94.
Perforators, different shapes of, 12, 13.
Perforators of bone, 63.
   " stone, 13.
Perry's Expedition to Japan, 8.
Pestles, 41, 42.
   " for preparing food, 43.
Pierced tablets. 29, 33.
Pin-shaped articles made of marine univalves, 68.
Pipes, 45, 47, 49.
Plates of stone. 37.
Plummets, 26.
Porto Rico, collections from, 4.
Potstone, material for vessels, 36.
Pottery, 73.
   " painted and decorated, 79, 81.
Powell, Major J. W., 2, 40.

(108)

INDEX.

Bottles made of shells, 69.
" of clay, 85.
Relics, prehistoric, from Europe, 4.
" of bone and horn from California and New York, 63.
Santa Barbara Islands, 31, 63, 68.
Schoolcraft, Mr., 33, 43.
Schumacher, Mr. Paul, 31, 37, 51, 63, 70, 72, 91.
Scraper-like implements, 23, 26.
Scrapers, different shapes of, 13.
" with handles, 13.
Sculptures, 54, 55, 57.
Serpentine, material for vessels, 27.
Shell-money, 70.
Shells used for ornament, 51, 52, 60, 69.
Shells, utensils of, 68, 67.
Silver, traces of, 60.
Smith, Captain John, 50.
Smithsonian Contributions to Knowledge, 45.
Spade-like implements, 25, 26.
Spear-heads, different shapes of, 11, 12.
" resemble large arrow-heads, 10.
Spear-heads, so-called, used with handles, 10.
Specimens from Central and South America, 4.
Spindle whorls, Mexican, 87.
Squier, Mr. E. G., 4.

Squier & Davis, Messrs., 45, 46, 48.
Statuettes, Mexican, 88.
Stevens, Mr. E. T., 45.
Stone as raw material, 7.
" flaked and chipped, 7.
" implements, 90.
" " their use, often doubtful, 2.
Stone knives for cutting leather, 9.
Stone, pecked, ground and polished, 17.
Stone plates, 27.
" vessels, 30.
Stones, club-head-shaped, 31.
" discoidal, used in Indian games, 29, 32.
Stones, egg-shaped, 28.
" used in grinding and polishing, 34, 35.
Stripod slate, material for ornaments, 53.
System adopted in arranging the Smithsonian collection relating to North American Ethnology, 97.

Tablets, pierced, 32, 33.
Teeth, perforated for ornaments, 64.
Terra-cotta figures, 84.
Tilhuggersteens, 22.
Tissues and implements from China and Japan, 5.
Tomahawks, 17.
Tools used by modern Indians, 23.
Torquemada, 8.

Toy canoes, 88.
Trichecus manatus, 16.
Tubes, 43, 41.
" ornaments or amulets, 43.
Tylor, Mr. E. B., 9.
Typical objects only described, 3.
Unalaska, ice-chisel from, 16.
Utensils made of shells, 67.
Vases, Mexican, 62, 81.
Veeргая, 44.
Verazzano, 50.
Vessels of clay for culinary purposes, 73.
" stone, 30.
Wampum-beads, 69.
Weapons of parade, drilled, 23, 24.
Weapons, utensils, etc. from Asia, Africa, Australia, etc., 4.
Wedges or Celts, 17.
Weights for fishing-lines, 26.
Whipple, Lieut., 81.
Whistles of bird-bones, 63.
" clay, 85.
Whittlesey, Mr. Charles, 56.
Wilkes, Capt., 5.
Williams, Roger, 62, 58, 76.
Wood, manufactures of, 66.
Wyman, Prof. Jeffries, 71.

Yarrow, Dr. H. C., 31.

www.ingramcontent.com/pod-product-compliance
Lightning Source LLC
Chambersburg PA
CBHW031356160426
43196CB00007B/829